Physics
A Contemporary Perspective

Student Workbook
Volume One

Preliminary Edition

Physics
A Contemporary Perspective

Student Workbook
Volume One

Preliminary Edition

Randall D. Knight
California Polytechnic State University - San Luis Obispo

ADDISON-WESLEY

An imprint of Addison Wesley Longman, Inc.

Reading, Massachusetts • Menlo Park, California • New York • Harlow, England
Don Mills, Ontario • Sydney • Mexico City • Madrid • Amsterdam

Reproduced by Addison Wesley Longman from camera-ready copy supplied by the author.

Copyright © 1997 Addison Wesley Longman, Inc.

All rights reserved. No part of this publication may be reproduced, stored in a retrieval system, or transmitted, in any form or by any means, electronic, mechanical, photocopying, recording, or otherwise, without the prior written permission of the publisher. Printed in the United States of America.

ISBN   0-201-43166-1

1 2 3 4 5 6 7 8 9 10 CRS 99989796

Table of Contents

	Introduction	
1.	Concepts of Motion	1
2.	Vectors and Coordinate Systems	13
3.	Kinematics: The Mathematics of Motion	19
4.	Force and Motion	39
5.	Dynamics I: Newton's Second Law	47
6.	Dynamics II: Motion in a Plane	57
7.	Dynamics III: Newton's Third Law	67
8.	Momentum and Its Conservation	81
9.	Concepts of Energy I: Work and Energy	89
10.	Concepts of Energy II: Potential Energy and Conservation	103
11.	Expanding the Concept of Energy	111
12.	Newton's Theory of Gravity	117
13.	Oscillations	121
14.	Traveling Waves	135
15.	Standing Waves	151
16.	Interference	161
17.	Diffraction	173
18.	A Closer Look at Light and Matter	177
	Dynamics Worksheets	
	Conservation Worksheets	

Introduction

A Physics Workbook

A Physics Workbook is a companion to the text *Physics: A Contemporary Perspective*. This workbook consists of exercises that give you an opportunity to practice the ideas and techniques presented in the text and in class. These exercises are intended for you to do on a daily basis, right after the topics have been discussed in class and are still fresh in your mind. Learning physics, as in learning any skill, requires regular practice of the basic techniques, and that is what this workbook is all about. Successful completion of the workbook exercises will prepare you to tackle the more challenging end-of-chapter homework problems in the text.

You will find that nearly all of the exercises are *qualitative* rather than *quantitative*. They ask you to draw pictures, interpret graphs, write short explanations, or provide other answers that do not involve calculations or numbers. The purpose of these exercises is to give you practice with the basic thinking tools you need *before* to attempting quantitative problems. It is highly recommended that you do these exercises prior to attempting the end-of-chapter problems

The exercises in this workbook are keyed to *specific sections* of the text. They assume that you have read that section, and they will give you a chance to practice the new ideas introduced in that section. You should keep the text beside you as you work and refer to it often. You will usually find guidelines, figures, or examples in the text that are directly relevant to the exercises. When asked to draw figures or diagrams, you should attempt to draw them so that they look much like the figures and diagrams in the text.

Since the exercises relate to specific sections the text, you should answer them on the basis of information presented in *just* that section. You may have learned new information in Section 7 of a chapter, but you should not use that when answering exercises from Section 4. There will be ample opportunity, when you reach the Section 7 exercises, to use that information there.

You will need a few "tools" to complete these exercises. Many of the exercises will ask you to *color code* your answers by drawing some items in black, others in red, and perhaps yet others in blue. You will need to purchase a few colored pencils to do this. The author highly recommends that you work in pencil, rather than ink, so as to facilitate erasures and corrections. Few are the individuals who make so few mistakes as to be able to work in ink! In addition, you'll find that a small six-inch ruler, which is easy to carry, will come in handy for drawings and graphs.

You will find, as the year goes along, that physics is a way of *thinking* about how the world works and why things happen as they do. We will be interested primarily in finding relationships and seeking explanations, only secondarily in computing numerical answers. In many ways, the "thinking tools" developed in this workbook are what the course is all about. If you take the time to do these exercises regularly and to review the answers, in whatever form your instructor provides them, you will be well on your way to success in physics.

To the instructor: The exercises in this workbook can be used in many ways. You can have students work on some of the exercises in class, in small groups, as part of an active-learning strategy. Or you can do the same in recitation sections or laboratories. This approach allows you to discuss the answers immediately, to answer student questions, and the improvise follow-up exercises when needed.

Alternatively, the exercises can be assigned as homework. The pages are perforated for easy tear-out, and the page breaks are in logical places so that you can assign the sections of a chapter that you would likely cover in one day of class. Exercises should be assigned at the end of one class, for the sections covered that day, and due at the beginning of the next class. Collecting them at the beginning of class, then going over two or three that are likely to cause difficulty, is an effective means of quickly reviewing major concepts from the previous class and launching a new discussion.

If used as homework, it is *essential* for students to receive *prompt* feedback. Ideally this would occur by having the exercises graded, with written comments, and returned at the next class meeting. Posting fairly detailed answers immediately after class also works. Lack of prompt feedback can negate much of the value of these exercises. Placing similar qualitative/graphical questions on quizzes and exams, and telling students at the beginning of the term that you will do so, encourages students to take the exercises seriously and to check the answers.

The author has been successful with assigning *all* exercises in the workbook as homework, collecting and grading them every day through Chapter 4, then collecting and grading them on about one-third of subsequent days on a random basis. Student feedback, from end-of-term questionnaires, reveals three prevalent attitudes toward the workbook exercises:

i. They think it is an unreasonable amount of work.
ii. They agree that the assignments force them to keep up and not get behind.
iii. They recognize, by the end of the term, that this is an extremely valuable learning tool.

However you choose to use these exercises, they will significantly strengthen your students' conceptual understanding of physics.

Following the workbook exercises are optional Dynamics Worksheets and Conservation Worksheets for use with end-of-chapter problems. Their use is recommended to help students acquire good problem solving habits. Problems in the text marked with the icon ✐ are intended to be done on worksheets. For Chapter 1 problems, which don't use the lower half of the worksheet, you can have students do two problems per worksheet with the second placed in the mathematical model section.

Chapter 1

Concepts of Motion

1.1   **Introduction**

1.2   **Basic Types of Motion**

1.3   **Motion Diagrams**

Using the particle model, draw motion diagrams for each motion described below. Number the positions in order, as shown in Fig. 1-8 in the text. Be neat and accurate!

1. A car accelerates forward from a stop sign and eventually reaches a steady cruising speed of 45 miles per hour.

2. An elevator starts from rest at the 100th floor of the Empire State Building and descends, with no intermediate stops, until coming to rest on the ground floor. (Draw this one *vertically* since the motion is vertical.)

3. Suzy Skier starts *from rest* at the top of a 30° snow-covered slope and skies to the bottom. (Orient your diagram correctly, as seen from the *side*, and label the 30° angle.)

2   Chapter 1 Concepts of Motion

4. The Space Shuttle orbits the earth in a circular orbit, completing one revolution each 90 minutes.

5. Bob throws a ball at an upward 45° angle from a third story balcony. The ball lands on the ground below.

Several motion diagrams are shown below. For each, write a short description of the motion of an object that will match the diagram. Your descriptions should name *specific* objects and be phrased similarly to the descriptions of Exercises 1 - 5. (Note the axis labels on 8 and 9.)

6.

```
                                 stops
 •            •        •      •   • •
 1            2        3      4   5 6
```

8.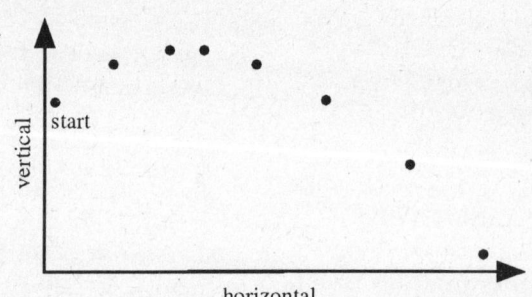

7.
```
1 •  starts
2 •
3 •

4 •



5 •
```

9.

1.4 Position and Time

10. The figure below shows the location of an object at three successive times: 1, 2 and 3.

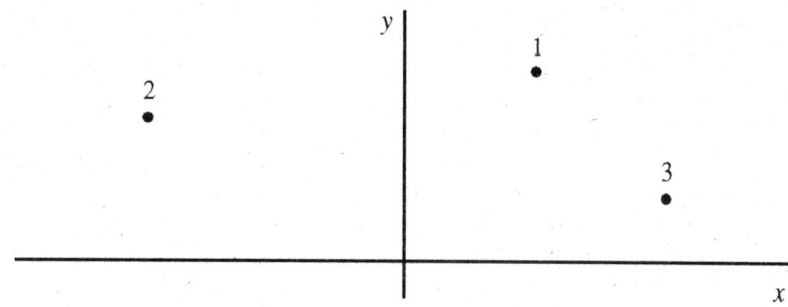

a) Use a red pencil to draw *and label*, on the figure, the three position vectors $\vec{r}_1$, $\vec{r}_2$, and $\vec{r}_3$ at times 1, 2, and 3.
b) Use a blue or green pencil to draw the object's "trajectory" $1 \to 2 \to 3$.
c) Use a black pencil to draw the displacement vector $\Delta \vec{r}$ from the initial to the final position.

11. In Exercise 10, is the object's displacement equal to the distance the object travels? Explain.

12. Redraw your five motion diagrams of Exercises 1 - 5 in the space below, but now add and label the displacement vectors $\Delta \vec{r}$ on the diagrams.

1.5 Velocity

13. The figure below shows the positions of a moving object in two successive frames of film. Frame 1 occurs prior to frame 2.

2 •

• 1

a) Use a black pencil to draw *and label* the velocity vector corresponding to these two frames.

b) Establish a coordinate system of your choice by drawing a set of *xy*-axes. Place the origin wherever you wish. Then use a red pencil to draw *and label* the position vectors $\vec{r}_1$ and $\vec{r}_2$.

c) Does $\vec{r}_2 = \vec{r}_1 + \vec{v}$ for the three vectors on *your diagram*? Explain.

Exercises 14 - 20: Draw motion diagrams for each of the following motions. Use the particle model. Show and label the *velocity* vectors.

14. A rocket-powered car on a test track accelerates from rest to a high speed, then coasts at constant speed after running out of fuel. Use five or six points for each part of the motion. Draw a dotted line across your diagram to indicate the point at which the car runs out of fuel.

15. Galileo drops a ball from the Leaning Tower of Pisa. Consider the ball's motion from the moment it leaves his hand until a microsecond before it hits the ground. Your diagram should be vertical. Use five or six points.

16. An elevator starts from rest at the ground floor. It accelerates upward for a short time, then moves with constant speed, and finally brakes to a halt at the tenth floor. Draw dotted lines across your diagram to indicate where the acceleration stops and where the braking begins. You'll need ten or twelve points to indicate the motion clearly.

17. A bowling ball being returned from the pin area to the bowler rolls at a constant speed. It then goes up a ramp and exits onto a level section at very low speed. You'll need ten or twelve points to indicate the motion clearly.

18. Tommy Trackstar runs one time around a running track at constant speed. The track has straight sides and semicircular ends. Use a bird's-eye view looking down on the track. Include about 20 points on your motion diagram.

19. A car is parked on a hill. The brakes fail, and the car rolls down the hill with an ever increasing speed. At the bottom of the hill it runs into a hedge and comes to a halt.

20. Andy is standing on the street and Bob is standing on the second-floor balcony of their apartment. Andy throws a baseball up to Bob. Consider the ball's motion from the moment it leaves Andy's hand until a microsecond before Bob catches it.

1.6 Acceleration

Note: Beginning with this section, and for future motion diagrams, you will "color code" the vectors. Draw velocity vectors with a *black* pencil and acceleration vectors with a *red* pencil.

Exercises 21 - 26: The figures below show an object's position in three successive frames of film. The object is moving in the direction $1 \rightarrow 2 \rightarrow 3$. For each diagram:

a) Draw *and label* (as $\vec{v}_i$ and $\vec{v}_f$) the initial and final velocity vectors.
b) Below the motion diagram, redraw the two velocity vectors with their tails together. This will look similar to Figure 1-21b in the text.
c) Use vector subtraction to determine the acceleration $\vec{a}$ from $\vec{v}_i$ and $\vec{v}_f$.
d) Draw and label $\vec{a}$ at the proper location on the original motion diagram, similar to Fig. 1-21c.
e) Decide whether the object is speeding up, slowing down, or moving at a constant speed. Write your answer beside the diagram.

Chapter 1 Concepts of Motion

Exercises 27 - 34: Draw a complete motion diagrams for each of the following. Include and label both velocity vectors $\vec{v}$ and acceleration vectors $\vec{a}$. Color code them appropriately.

27. Galileo drops a ball from the Leaning Tower of Pisa. Consider its motion from the moment it leaves his hand until a microsecond before it hits the ground.

28. Betty is driving her car at a steady 30 mph when a small furry creature runs into the road in front of her. She hits the brakes and skids to a stop. Consider her motion from 2 seconds before she starts braking until she comes to a complete stop.

29. A ball is rolled up a smooth board tilted at a 30° angle, and then it rolls back to its starting position.

30. A bowling ball being returned from the pin area to the bowler rolls at a constant speed, then up a ramp, and finally exits onto a level section at very low speed.

31. Tommy Trackstar runs one time around a running track at constant speed. The track has straight sides and semicircular ends. Use a bird's-eye view looking down on the track.

32. A cannon ball is fired from a Civil War cannon up onto a high cliff. Consider the cannon ball from the moment it leaves the cannon until a microsecond before it hits the ground.

33. A plane flying north at 300 mph turns slowly to the west without changing speed. Draw the motion diagram from a viewpoint above the plane.

34. Two sprinters, Diane and Debbie, start side by side. Diane has only reached the 80-meter point when Debbie crosses the finish line of the 100-meter dash.

1.7 From Words to Symbols

35. The motion diagrams below show an initial point 0 and a final point 1. A pictorial model would define five symbols: x_0 and x_1 (or y_0 and y_1), v_0 and v_1, and a. Determine whether each of these quantities is positive, negative, or zero. Place either +, −, or 0 in each cell of the table below.

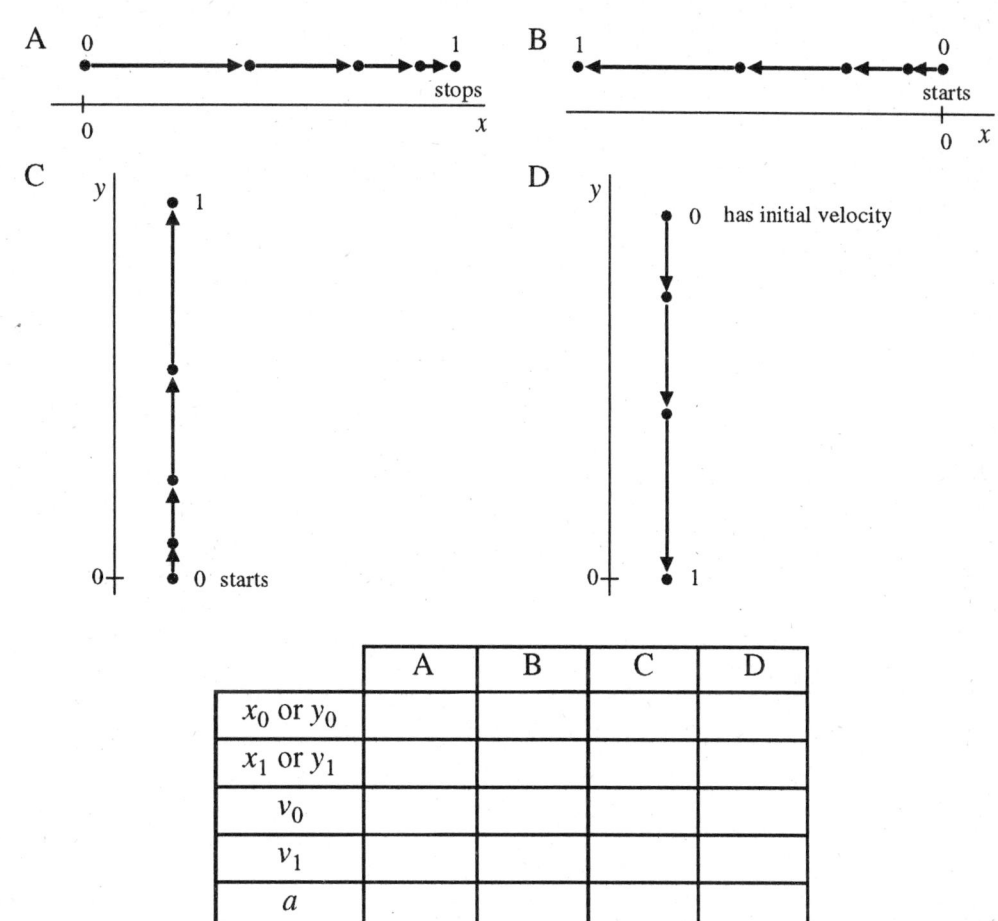

36. The three symbols x, v, and a have eight possible combinations of *signs*. For example, one combination is $(x, v, a) = (+, -, +)$.

a) List all eight combinations of signs for x, v, a:

#	combination	#	combination
1	_____	5	_____
2	_____	6	_____
3	_____	7	_____
4	_____	8	_____

b) For each of the eight combinations of signs you identified in Part a), draw a complete motion diagram showing the motion of an object that has these signs for x, v, and a. Draw the diagram *below* the axis whose number corresponds to Part a), and use labeled and color-coded $\vec{v}$ and $\vec{a}$ vectors.

#1
$x = 0$, x

#2
$x = 0$, x

#3
$x = 0$, x

#4
$x = 0$, x

#5
$x = 0$, x

#6
$x = 0$, x

#7
$x = 0$, x

#8
$x = 0$, x

1.8 Working with Motion Diagrams and Pictorial Models

No exercises.

Chapter 2

Vectors and Coordinate Systems

2.1 Scalars and Vectors

2.2 Properties of Vectors

Find and label the vector sum $\vec{A} + \vec{B}$ for the vectors shown below.

1.
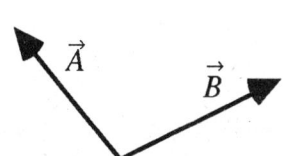

2.
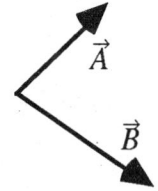

3.
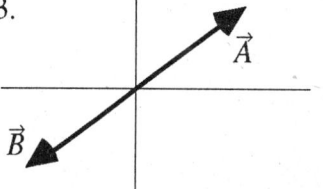

4. Use a figure to show that vector addition is associative — that is, that
$$(\vec{A} + \vec{B}) + \vec{C} = \vec{A} + (\vec{B} + \vec{C}).$$

Find and label the vector difference $\vec{A} - \vec{B}$ for the vectors shown below.

5.
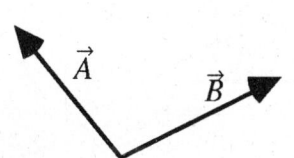

6.
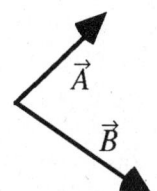

7.
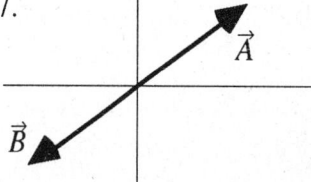

14   Chapter 2 Vectors and Coordinate Systems

8. Draw and label the vector $2\vec{A}$ and the vector $\frac{1}{2}\vec{A}$.

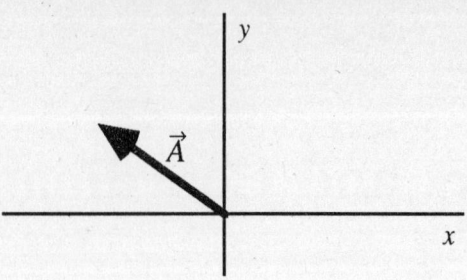

9. Is it possible to *add* a scalar to a vector? If so, demonstrate. If not, explain why.

10. How would you define the *zero vector* $\vec{Z} = 0$?

2.3   Component Vectors and Unit Vectors

Draw and label the *x*- and *y*-component vectors for each of the following vectors:

11.   12.   13.

Write the following vectors in component form (e.g. $3\hat{i} + 2\hat{j}$).

14. 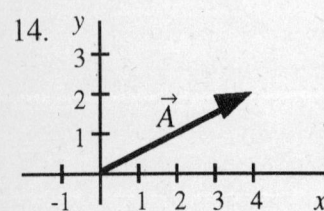   15. 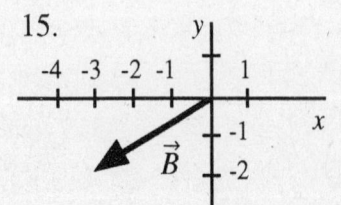   16.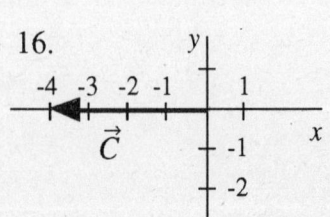

Chapter 2 Vectors and Coordinate Systems   15

17. What is the vector sum $\vec{A} + \vec{B} + \vec{C}$ of the three vectors defined in Exercises 14 - 16? Write your answer in *component* form.

Draw and label the following vectors on the axes below:

18. $-\hat{i} + 2\hat{j}$

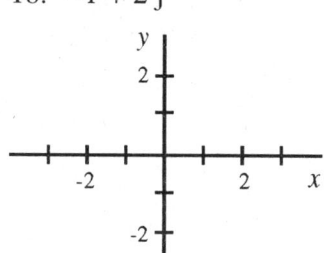

19. $-2\hat{j}$

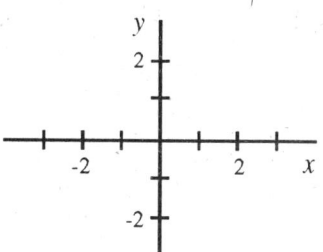

20. $3\hat{i} - 2\hat{j}$

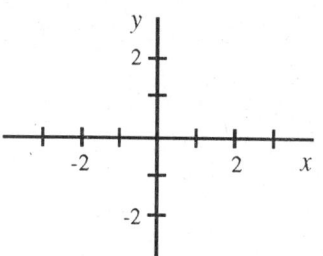

21. Using the idea of components, how would you define the *zero vector* $\vec{Z} = 0$?

Determine the numerical values of the *x*- and *y*-components of these vectors:

22.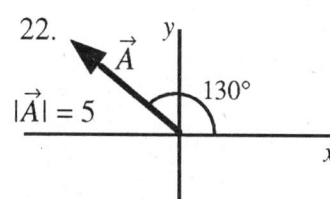

$A_x =$ _____

$A_y =$ _____

23.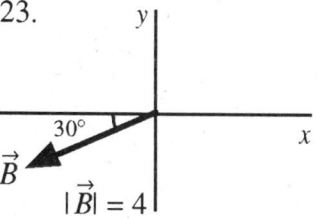

$B_x =$ _____

$B_y =$ _____

24.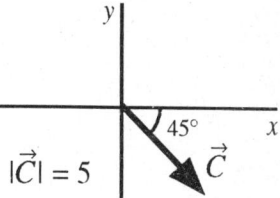

$C_x =$ _____

$C_y =$ _____

Chapter 2 Vectors and Coordinate Systems

For each of the three vectors given below:
  a) Draw the vector on the axes provided.
  b) Identify and label an angle θ to describe the direction of the vector.
  c) Find the magnitude and the angle of the vector.

25. $\vec{A} = 2\hat{i} + 2\hat{j}$

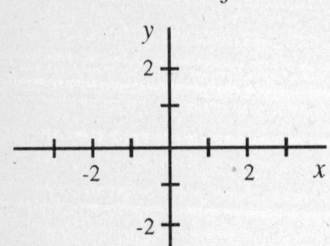

$|\vec{A}| =$ _____

$\theta =$ _____

26. $\vec{B} = -2\hat{i} + 2\hat{j}$

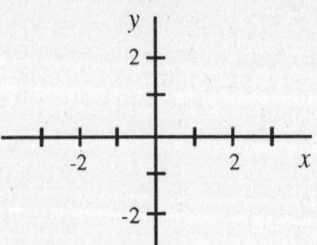

$|\vec{B}| =$ _____

$\theta =$ _____

27. $\vec{C} = -3\hat{i} - 2\hat{j}$

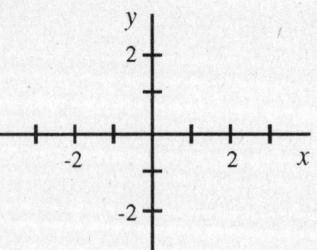

$|\vec{C}| =$ _____

$\theta =$ _____

28. Can a vector have a component equal to zero and still have nonzero magnitude? Explain.

29. Can a vector have zero magnitude if one of its components is nonzero? Explain.

30. Suppose two vectors have unequal magnitudes. Can their sum be zero? Explain.

Vector $\vec{A}$ is defined as $\vec{A} = (5, 30°$ above the horizontal). Determine the components A_x and A_y in the three coordinate systems shown below. Show your work below the figure.

31.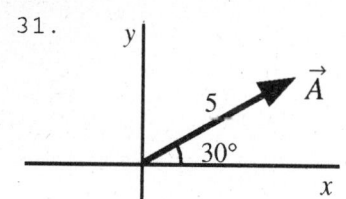

$A_x = $ _____

$A_y = $ _____

32.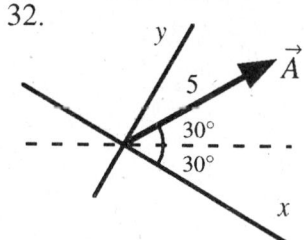

$A_x = $ _____

$A_y = $ _____

33.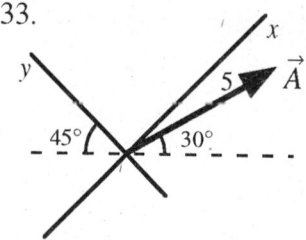

$A_x = $ _____

$A_y = $ _____

2.4 Significant Figures

34. How many significant figures does each of the following numbers have?

a) 6.21 _____   e) 62100. _____   i) .62 _____

b) 62.1 _____   f) 0.0621 _____   j) 1.0621 _____

c) 6210 _____   g) 0.620 _____   k) 6.21×10^3 _____

d) 62100 _____   h) 0.62 _____   l) 62.1×10^3 _____

35. Compute the following numbers, applying the significant figure standard adopted for this text.

a) $33.3 \times 25.4 =$ _____   d) $33.3 \times 45.1 =$ _____

b) $33.3 - 25.4 =$ _____   e) $33.3^2 =$ _____

c) $33.3 \div 25.4 =$ _____   f) $\sqrt{33.3} =$ _____

Chapter 3

Kinematics: The Mathematics of Motion

3.1 Introduction

3.2 Measurements and Units

1. Convert the following to SI units:

   a) 9.12 μs   b) 3.42 km   c) 44 cm/ms   d) 80 km/hour
   e) 250 cm^3 f) 8 inches   g) 14 square inches   h) 60 miles/hour
   (Note: Think carefully about e) and g). A geometric diagram may help.)

20   Chapter 3 Kinematics: The Mathematics of Motion

2. Use Table 3-3 to assess whether or not the following statements are *reasonable*. (Note the comments on *assessment* and *reasonableness* at the end of Section 3.2.)

a) Joe is 180 cm tall.

b) I rode my bike to campus at a speed of 50 m/s.

c) Sammy Skier reaches the bottom of the slope going 25 m/s.

d) I can throw a ball 2 km.

e) I can throw a ball with a speed of 100 km/hour.

3. Justify the assertion that 1 m/s ≈ 2 mph by *exactly* converting 1 m/s to English units. By what percentage is this rough conversion in error?

3.3   Motion in One Dimension

4. Sketch position-versus-time graphs for the following motions. Include a numerical scale on both axes, with units that are *reasonable* for this motion. Some numerical information is given in the problem, but for other quantities make *reasonable estimates*. (Note: A *sketched* graph simply means hand-drawn, rather than carefully measured and laid out with a ruler. But a sketch should still be neat and as accurate as is feasible by hand. It also should include labeled axes and, if appropriate, it should have tick-marks and numerical scales along the axes.)

a) A student walks to the bus stop, waits for the bus, then rides to campus. Assume that all the motion is along a straight street.

b) A student walks slowly to the bus stop, realizes he forgot his paper that is due, and *quickly* walks home to get it.

c) The quarterback drops back 10 yards from the line of scrimmage then throws a pass 20 yards to the tight end, who catches it and sprints 20 yards to the goal. Draw your graph for the *football*. Think carefully about what the *slopes* of the lines should be.

d) Follow the horizontal motion of a tennis ball through several volleys back and forth. Let the net be at $x = 0$ and the server be on the left side of the net.

5. Interpret the following position-versus-time graphs by writing a very short "story" of what is happening. Be creative! Have characters and situations! Simply saying that "a car moves 100 meters to the right" doesn't qualify as a story. Your stories should make *specific reference* to information you obtain from the graphs, such as distances moved or time elapsed.

a) Moving car.

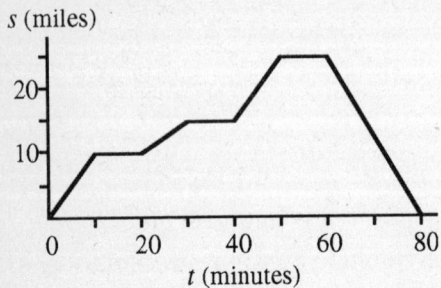

b) Sprinter.

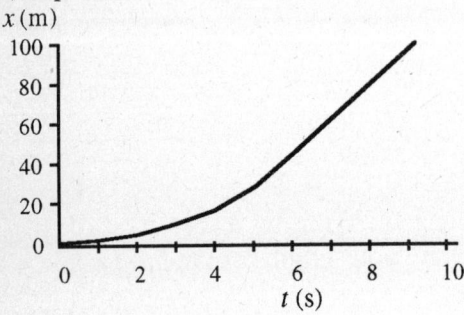

c) Submarine.

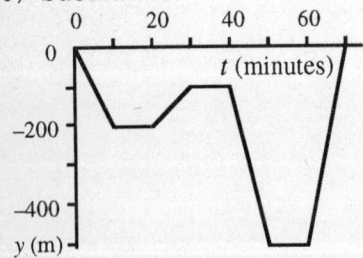

d) Two football players.

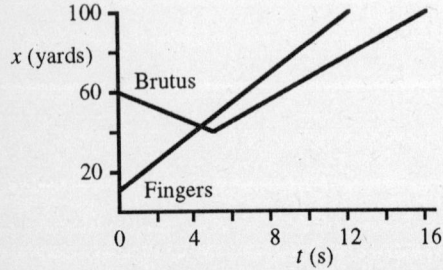

6. Consider this position-versus-time graph, which shows a *single* curve. Can you give this graph an interpretation? If so, then do so. If not, why not?

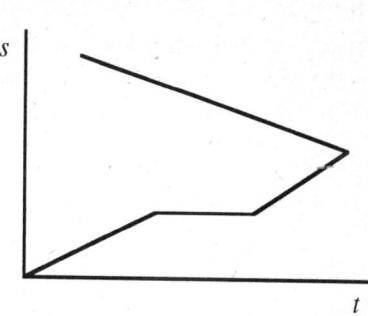

3.4 Uniform Motion

7. Sketch position-versus-time graphs for the following motions. Include appropriate numerical scales along both axes. A small amount of computation is necessary.

a) A parachutist open her parachute at an altitude of 1500 m, and she then descends slowly to earth at a steady speed of 5 m/s. Start your graph as her parachute opens.

b) Trucker Bob starts the day 120 miles west of Denver. He drives eastward for 3 hours at a steady 60 miles/hour before stopping for his coffee break. Let Denver be located at $x = 0$ and assume that the x-axis points to the east.

c) Quarterback Bill passes the ball straight down field with a speed of 15 m/s. It is intercepted 45 m away by Linebacker Larry, who is running up field at 7.5 m/s. Larry carries the ball 60 m to score. Let $x = 0$ be the point where Bill throws the ball. Draw the graph for the *football*.

24   Chapter 3 Kinematics: The Mathematics of Motion

8. The figure shows a position-versus-time graph for the motion of two objects, A and B, which are moving along the same axis.

a) At the instant $t = 1$ s, is the speed of A greater than, less than, or equal to the speed of B? Explain your reasoning.

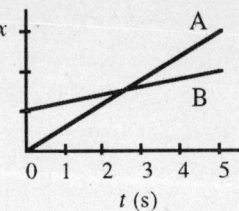

b) Do object's A and B ever have the *same* speed? If so, at what time or times? Explain your reasoning.

9. Interpret the following position-versus-time graphs by writing a very short "story" of what is happening. Your stories should make *specific reference* to the *speed* of the moving objects, which you will need to determine from the graphs. Assume that the motion takes place along a horizontal line.

(a)

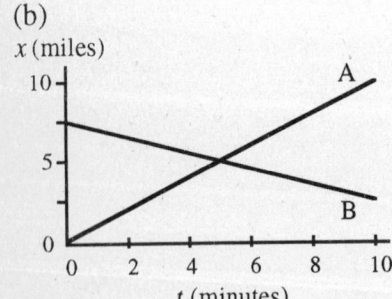

(b)

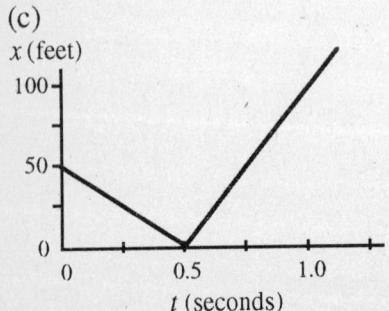

(c)

3.5 Instantaneous Velocity

10. Draw both a position-versus-time graph *and* a velocity-versus time graph for an object that is at rest at $x = 1$ m.

11. Below are eight position-versus-time graphs. For each, draw the corresponding velocity-versus-time graph directly below it. Both graphs should have the same time scale (i.e., a vertical line drawn through both graphs should connect the velocity v at time t with the position s at the *same* time t). There are no numbers in this question, but your graphs should indicate the *relative* speeds correctly.

a)
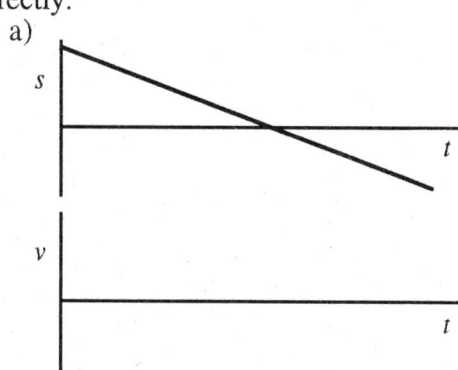

b)
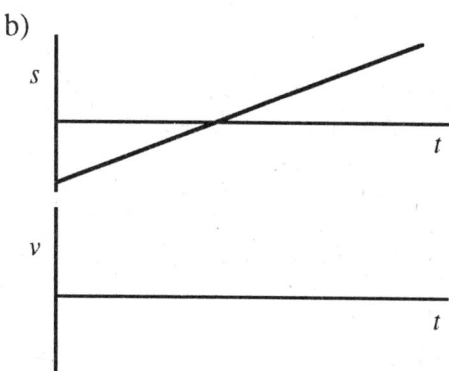

c)
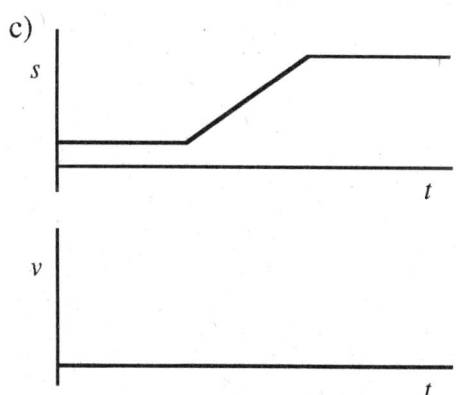

d)
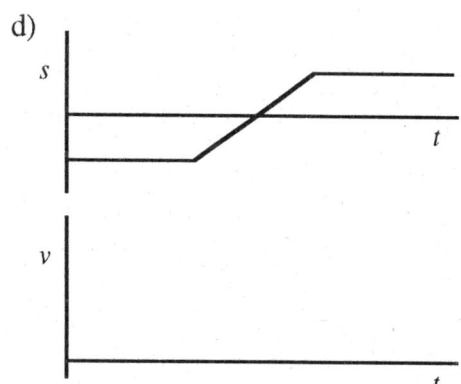

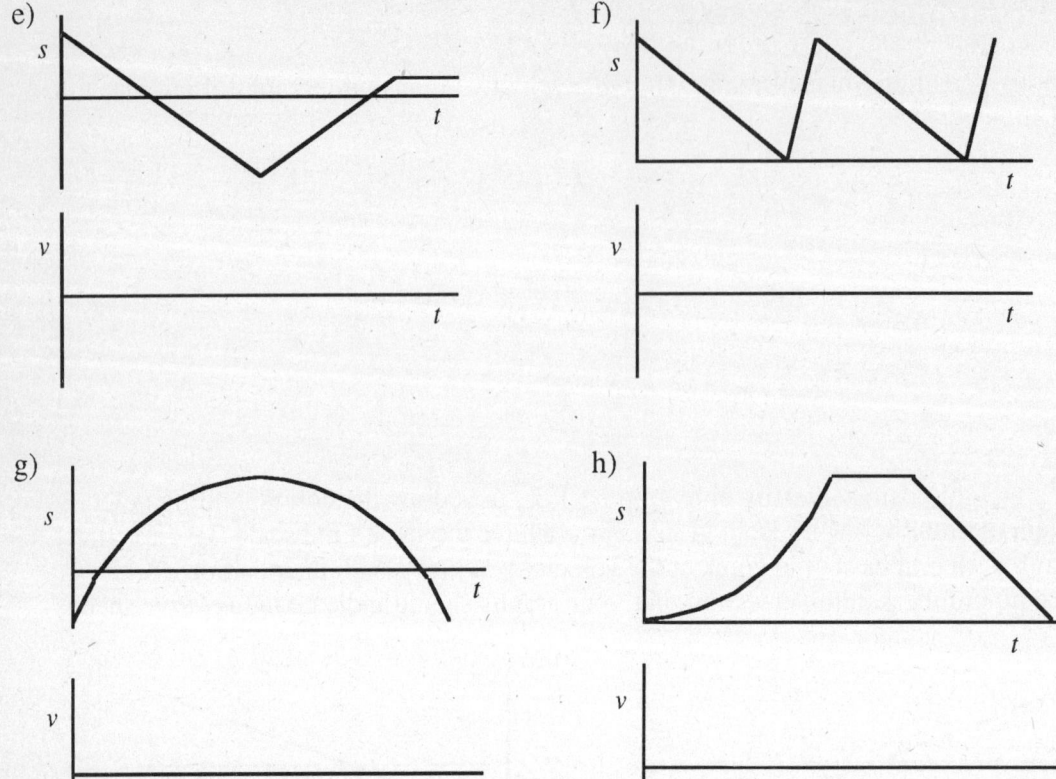

12. The figure shows a position-versus-time graph for the motion of two objects, A and B, which are moving along the same axis.

a) At the instant $t = 1$ s, is the speed of A greater than, less than, or equal to the speed of B? Explain your reasoning.

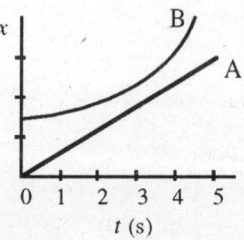

b) Do objects A and B ever have the *same* speed? If so, at what time or times? Explain your reasoning?

13. The figure shows a position-versus-time graph for a moving object. Various specific times are labeled A, B, ..., E. At which lettered point or points:
a) Is the motion slowest?

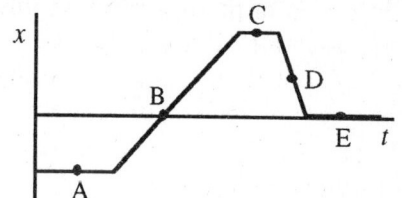

b) Is the motion the fastest?

c) Is the object moving at a constant non-zero velocity?

d) Is the object moving to the left?

e) Is the object turning around?

f) Is the object at rest?

14. The figure shows a position-versus-time graph for a moving object. Various specific times are labeled A, B, ..., F. At which lettered point or points:
a) Is the motion slowest?

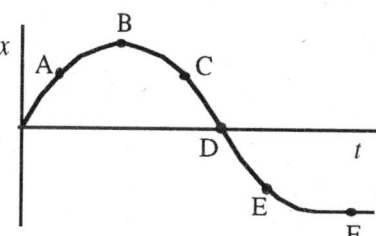

b) Is the motion the fastest?

c) Is the object moving at a constant non-zero velocity?

d) Is the object moving to the left?

e) Is the object speeding up?

f) Is the object slowing down?

g) Is the object turning around?

15. For each of the following motions, draw
    i) A motion diagram,
    ii) A position-versus-time graph, and
    iii) A velocity-versus-time graph.

a) A car starts from rest, steadily speeds up to 40 mph in 15 s, moves at a constant speed for 30 s, then brakes rapidly to a halt in 5 s.

b) A rock is dropped from a bridge and steadily speeds up as it falls. It is moving at 30 m/s when it hits the ground 3 s later. Let the *y*-axis point vertically up, and think carefully about the signs.

c) A pitcher winds up and throws a baseball with a speed of 40 m/s. One-half second later the batter hits a line drive with a speed of 60 m/s. The ball is caught 1 s after it is hit. From where you are sitting, the batter is to the right of the pitcher. Draw your motion diagram and graph for the *ball*.

16. The figure shows six frames from the motion diagram of two moving cars, A and B.

a) Describe, in words, what you would *see* if you watched this motion.

b) Draw both a position-versus-time graph and a velocity-versus-time graph. Show the motion of *both* cars on each graph, such as in Exercises 8 or 12, and label them A and B.

c) Do the two cars ever have the same position at one instant of time? If so: i) identify in which frame number, and ii) identify the point on your graphs of part b).

d) Do the two cars ever have the same velocity at one instant of time? If so: i) identify in which frame number, and ii) identify the point on your graphs of part b).

17. The figure shows six frames from the motion diagram of two moving cars, A and B.

a) Describe, in words, what you would *see* if you watched this motion.

b) Draw both a position-versus-time graph and a velocity-versus-time graph. Show the motion of both cars on each graph, such as in Exercises 8 or 12, and label them A and B.

c) Do the two cars ever have the same position at one instant of time? If so: i) identify in which frame number, and ii) identify the point on your graphs of part b).

d) Do the two cars ever have the same velocity at one instant of time? If so: i) identify in which frame number, and ii) identify the point on your graphs of part b).

18. As you drive along the highway at a steady speed of 60 mph, you notice another car following you. The other driver patiently maintains a constant distance behind you for awhile, but eventually he decides to pass you. At the moment when the front of his car is exactly even with the front of your car, and you turn your head to smile at him, do the two cars have equal positions (as measured along the road axis)? Equal velocity? Both? Neither? Explain your reasoning.

3.6 Relating Velocity To Position

19. Below are shown two velocity-versus-time graphs. For each:
    i) Draw the corresponding position-versus-time graph,
    ii) Draw a motion diagram, and
    iii) Give a verbal description of the motion.

Assume that the motion takes place along a horizontal line and that $x_0 = x(t=0) = 0$.

a)
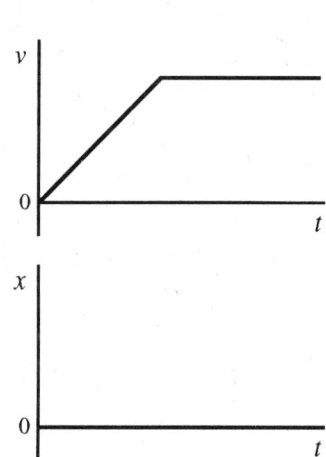

b)
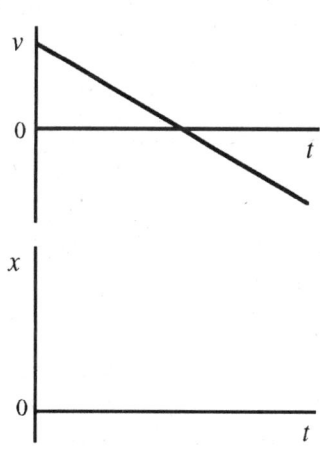

20. The figure shows the velocity-versus-time graph for a moving object whose initial position is $x_0 = x(t=0) = 20$ m. Find the object's position graphically, using the geometry of the graph, at the following times.

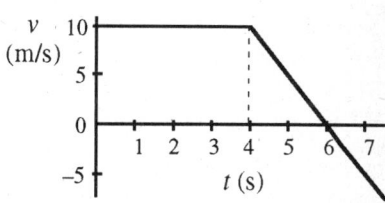

a) At $t = 3$ s.

b) At $t = 5$ s.

c) At $t = 7$ s.

d) You should have found a very simple relationship between your answers to b) and c). Can you explain this? What is the object doing?

21. Below are shown three velocity-versus-time graphs. Draw the corresponding position-versus-time graphs directly below them. For each, $x_0 = x(t = 0) = 0$.

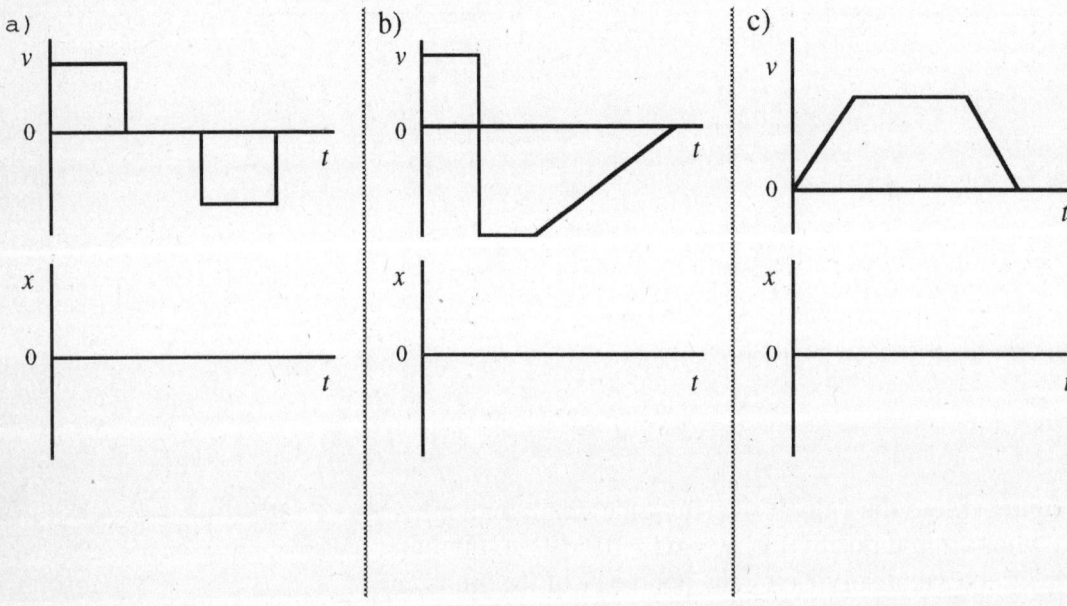

3.7 Constant Acceleration

22. A car is traveling north. Can its acceleration vector ever point south? Explain.

23. Give a specific example for each of the following situations. For each, provide:
    i) A description in words, and
    ii) A motion diagram.
 a) $a = 0$ but $v \neq 0$.

 b) $v = 0$ but $a \neq 0$.

 c) $v > 0$ and $a > 0$.

 d) $v > 0$ and $a < 0$.

24. Below are three velocity-versus-time graphs. For each:
    i) Draw the corresponding acceleration-versus-time graph.
    ii) Draw a motion diagram below the graphs.

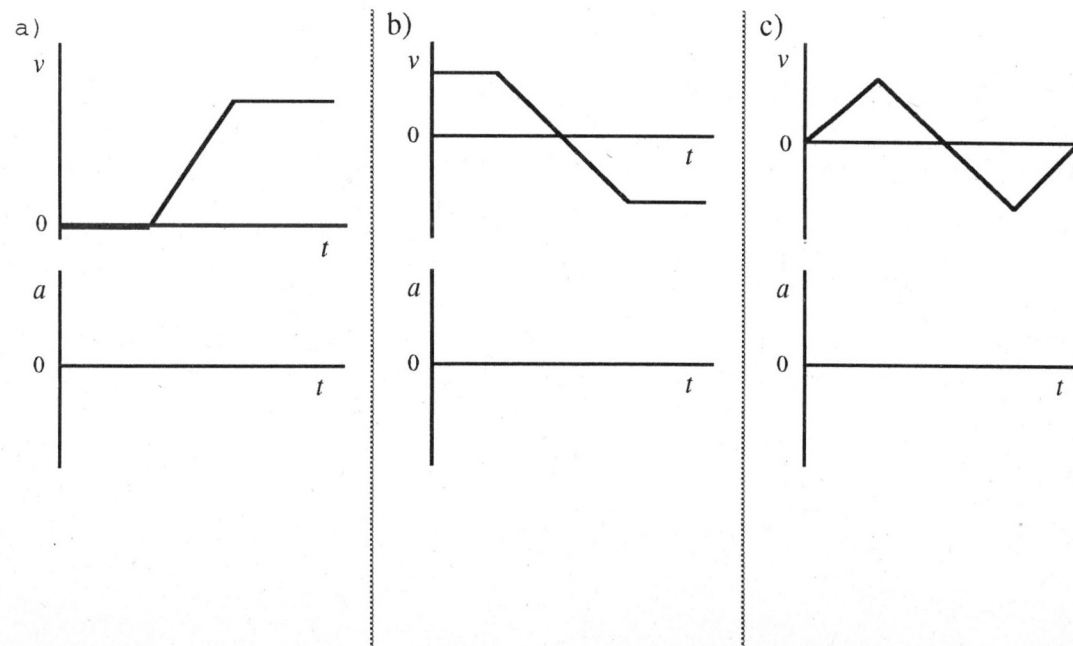

34 Chapter 3 Kinematics: The Mathematics of Motion

25. Below are three acceleration-versus-time curves. For each:
   i) Draw the corresponding velocity-versus-time curve, assuming that $v_0 = v(t = 0) = 0$.
   ii) Draw a motion diagram below the graphs.

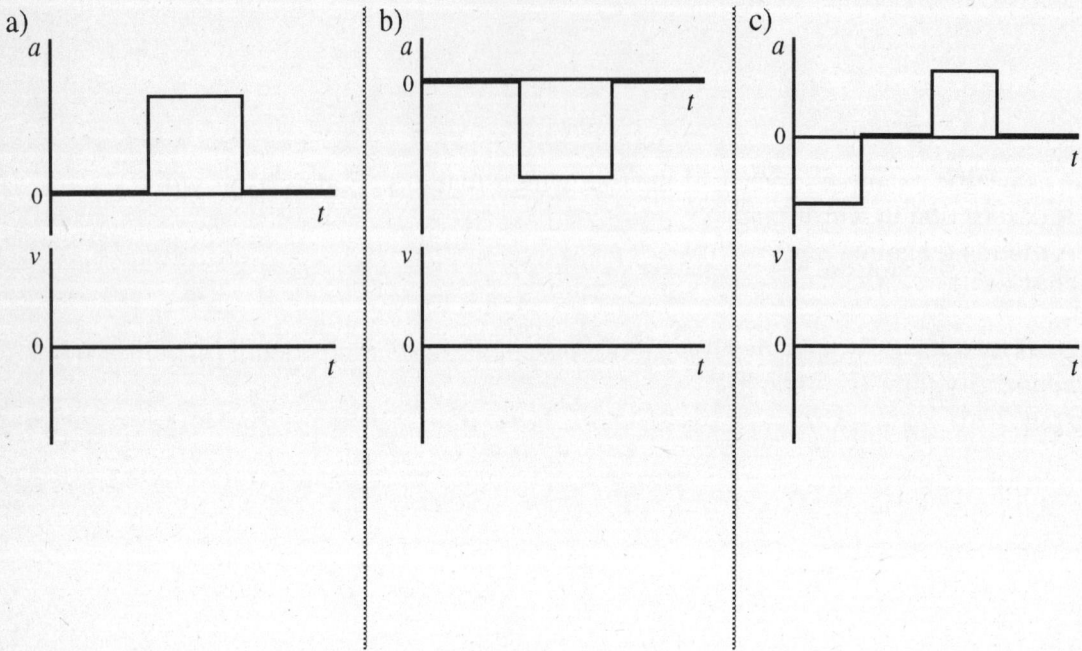

26. The figure below shows 9 frames from the motion diagram of two cars. Both cars begin to accelerate, with constant acceleration, in frame #4.

a) Which car has the largest initial velocity? How can you tell?

b) Which car has the largest final velocity? How can you tell?

c) Which car has the largest acceleration after frame #4? How can you tell?

d) Draw position, velocity, and acceleration graphs, showing both cars motion on each graph. This is a total of three graphs with two curves on each.

e) Do the cars ever have the same position at one instant of time? If so, identify in which frame or frames *and* identify it on your graphs?

f) Do the two cars ever have the same velocity at one instant of time? If so, identify the *two* frames between which this velocity occurs *and* identify it on your graphs.

g) Give a description, in words, of how these two cars move.

3.8 Instantaneous Acceleration

3.9 Free Fall and Inclined Planes

27. A ball rolling released from rest on an inclined plane has an acceleration of 2 m/s^2. Complete the table below showing its velocity at the times indicated. Do NOT use a calculator for this — this is a reasoning question, not a calculation problem.

time (s)	velocity (m/s)
0	0
1	
2	
3	
4	
5	
6	

28. A ball is thrown straight up into the air. What is the ball's acceleration
a) Just after leaving your hand?

b) At the very top (maximum height)?

c) Just before hitting the ground?

29. Alice throws a red ball straight up into the air, releasing it with velocity v_0. As she is throwing it, you happen to pass by in an elevator that is rising with constant velocity v_0. At the exact instant Alice releases her ball, you reach out of the elevator's window (this is a very fancy elevator!) and *gently* release a blue ball. Both balls are the same height above the ground at the moment they are released.

a) Describe, in words, the motion of the two balls as Alice sees them from the ground. In what ways are the motion of the red ball and the blue ball the same or different?

b) Describe the motion of the two balls as you see them from the moving elevator. In what ways are the motion the same or different?

c) Alice sees a definite "top" of the motion, where her red ball reaches a maximum height and then starts to fall. Call the time of maximum height t_1. As you watch from the elevator, do *you* see anything different or unusual about the red ball's motion at time t_1?

d) Does the red ball "stop" at time t_1 when Alice sees it at the very top of its trajectory? As part of answering this question, define what you mean by the word "stop."

30. A rock is *thrown* (not dropped) straight down from a bridge into the river below.
a) Immediately *after* being released, is the magnitude of the rock's acceleration greater than g, less than g, or equal to g. Explain your reasoning.

b) Immediately before hitting the water, is the magnitude of the rock's acceleration greater than g, less than g, or equal to g. Explain.

38 Chapter 3 Kinematics: The Mathematics of Motion

3.10 Putting It All Together

31. A bowling ball rolls along a level surface, then up a 30° slope, and finally exits onto another level surface at a much slower speed.

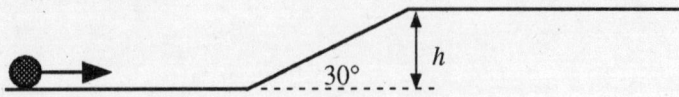

a) Draw position-, velocity-, and acceleration-versus time graphs for the ball.

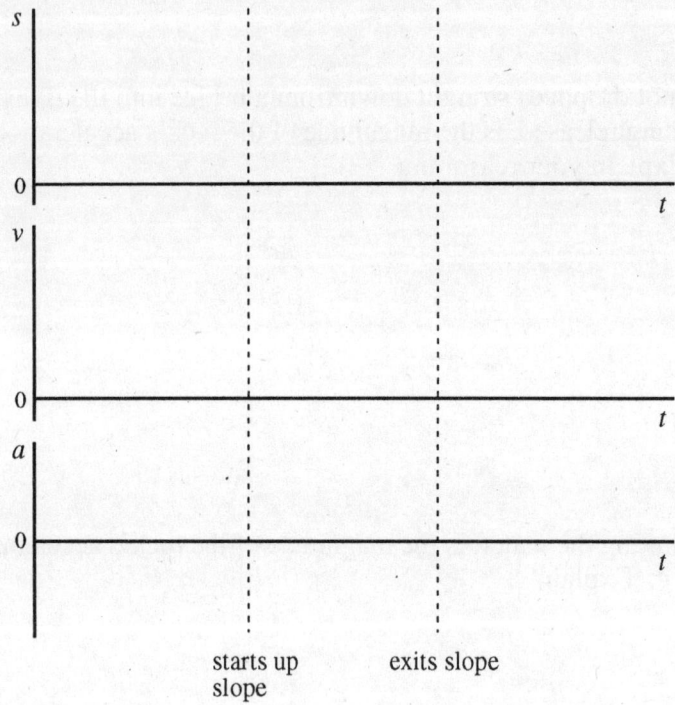

b) Suppose that the ball's initial speed is 5 m/s and its final speed is 1 m/s. Draw a pictorial model that you would use to determine the height h of the slope. Establish coordinate systems, define all symbols, list known information, and identify desired unknowns. (Don't actually solve the problem, just draw the complete pictorial model that you would use as a first step in solving the problem.)

Chapter 4

Force and Motion

4.1 Dynamics

4.2 Force

4.3 Force, Motion, and Newton's Second Law

1. The figure shows a force-versus-acceleration graph for a single object of mass m. Force and acceleration data have been plotted as individual points at several force strengths. A continuous line has been drawn through the points. Draw and label, directly on the figure, the force-versus-acceleration graphs for objects of mass
a) $2m$ b) $0.5m$

Use triangles Δ to show four points for the object of mass $2m$, then connect them with a continuous line. Use squares ❑ to show four points for the object of mass $0.5m$, then connect them with a continuous line.

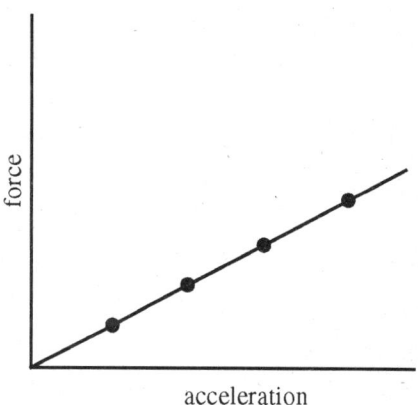

2. A constant force applied to object A produces an acceleration of 5 m/s². The same force applied to object B produces an acceleration of 3 m/s², and applied to object C it produces an acceleration of 8 m/s².

a) Which object has the largest mass? _____

b) Which object has the smallest mass? _____

c) What is the ratio of mass A to mass B (m_A/m_B)? _____

3. An object experiencing a constant force undergoes an acceleration of 10 m/s². What will be the acceleration of this object if:

a) The force is doubled? _____   d) The mass is doubled? _____

b) The force is tripled? _____   e) The mass is tripled? _____

c) The force is halved? _____   f) The mass is halved? _____

g) The force *and* the mass are both doubled? _____

h) The force is doubled *and* the mass is halved? _____

i) The force is halved *and* the mass is doubled? _____

40   Chapter 4 Force and Motion

4. Forces are shown on three objects below. For each:
a) Show and label the net force vector. Do this right on the figure.
b) Below the figure, show and label the object's acceleration vector.

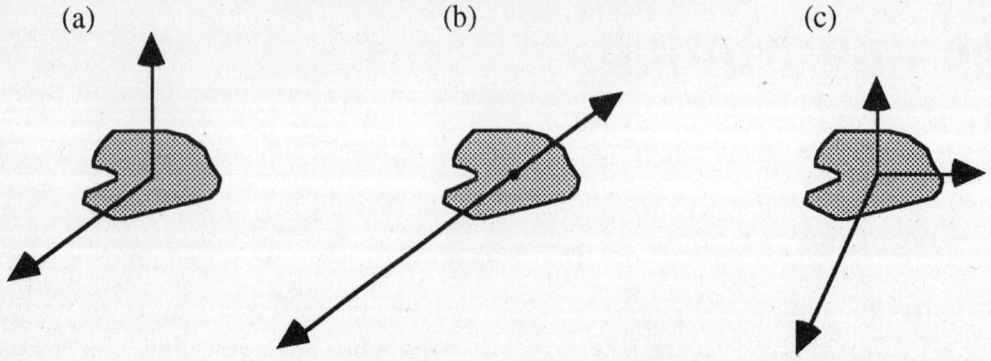

5. In the figures below, one force is missing. Use the given direction of acceleration to determine the missing force and show it on the figure. Do all work directly on the figure.

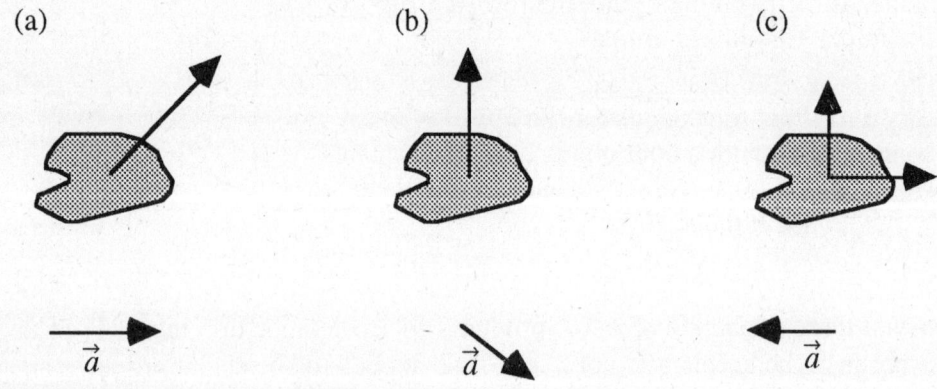

6. Below are four motion diagrams for a particle.
a) Explain *how* you would determine the direction of the net force on the particle at point 3.

b) Now do it. Draw and label the net force vector at point 3 with its tail attached to the point.

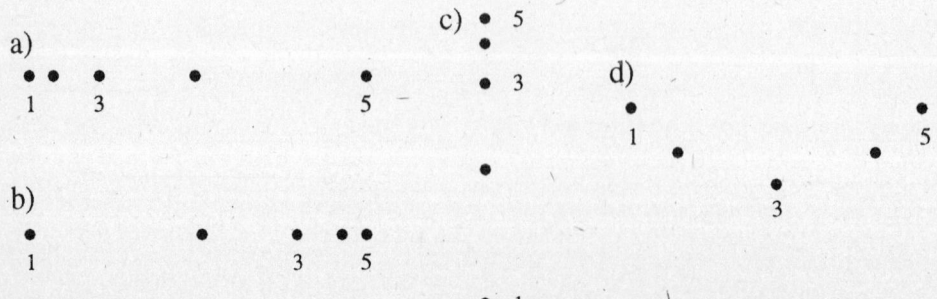

4.4 Inertia and Newton's First Law

7. If an object is at rest, can you conclude that there are no forces acting on it? Explain.

8. If a force is exerted on an object, is it possible for that object to be moving with constant velocity? Explain.

9. A hollow tube forms three-quarters of a circle. It is oriented in a vertical plane. A ball is shot through the tube at high speed. As the ball emerges from the other end, does it follow path A, path B, or path C? Explain your reasoning.

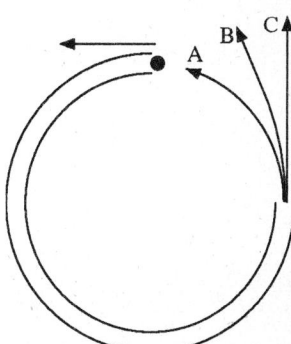

Chapter 4 Force and Motion

Note: The next two questions ask for a "physics explanation." A physics explanation should begin by stating a law or principle of physics and then proceed in a logical, step-by-step method to show how the phenomena is a consequence of the laws of physics. The explanation should be given in terms of specific concepts of physics, such as force, acceleration, velocity, and so on. When you are done, a reader of your explanation should be able to understand how and why the phenomena you are explaining is a logical outcome of the known laws of physics. Simple descriptions that do not refer to the laws and principles of physics are not physics explanations.

10. Give a physics explanation of why seatbelts are necessary in cars.

11. A well-known magician's "trick" is to jerk a tablecloth out from beneath a setting of dishes and glasses without upsetting them. Is this really a "trick?" Give a physics explanation.

4.5 A Short Catalog of Forces

No exercises.

4.6 Identifying Forces

Exercises 12 - 17: Follow the six-step procedure of this section to identify and name all forces acting on the object.

12. An elevator suspended by a cable is descending at constant velocity.

13. A car on a *very* slippery icy road is sliding head-first into a snow bank, where it gently comes to rest with no one injured.

14. A compressed spring is pushing a block across a rough horizontal table.

15. A brick is falling from the roof of a three-story building.

16. Block A and B are connected by a string passing over a pulley. Block B is falling and dragging block A across a frictionless table. Let block A be "the object" for analysis.

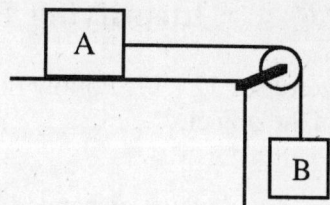

17. A jet plane is climbing at a 30° angle. Air resistance is not negligible.

4.7 Free-Body Diagrams

Exercises 18-25:
a) Draw a picture and identify the forces, then
b) Draw a complete free body diagram for the object, following each of the steps given in this section. Be sure to think carefully about the direction of $\vec{F}_{net}$.

NOTE: Color code free body diagrams to match the colors of motion diagrams. In these and in future exercises, draw *individual* force vectors with a *black* pencil and the *net* force vector $\vec{F}_{net}$ with a *red* pencil.

18. A heavy crate is being lifted straight up at constant speed by a steel cable.

Chapter 4 Force and Motion 45

19. A locomotive is *pushing* a boxcar along the rails at a steadily increasing speed. The friction of the rails is negligible. Let the boxcar be "the system" for analysis.

20. A boy is pushing a box across the floor at a steadily increasing speed. Friction is *not* negligible. Let the box be "the system" for analysis.

21. You are going to use a slingshot to shoot a rock straight up into the air. You've place the rock in the slingshot and pulled the rubber band down. Draw separate diagrams for the rock:
a) As you hold it, with the rubber band stretched, prior to its release.
b) One microsecond after you release it.

22. a) Your car ran out of gas while driving up a hill and is now *coasting* to a halt. Friction is small but not negligible.

b) The brakes failed, and now your car is rolling backwards down the hill.

23. Betty is skiing down a hill. She is still accelerating. Air resistance is *not* negligible.

24. Block B has just been released and is beginning to fall. Consider block A to be "the system."

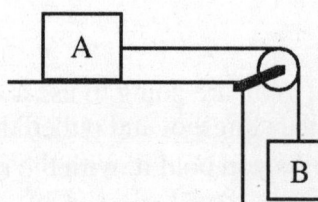

25. You flipped a coin to determine whether you or your brother has to wash the dishes. Draw separate diagrams for the coin: a) On its way up, and b) On its way back down. Air resistance is negligible.

Chapter 5

Dynamics I: Newton's Second Law

5.1 A Strategy for Force and Motion Problems

1. If you know all of the forces acting on a moving object, can you tell the direction it is moving? Give a physics explanation as to why or why not. (Review the comment on Workbook page 42 about "physics explanations.")

2. Write several sentences explaining why you agree or disagree with the statement: "Forces cause an object to move."

3. a) As an elevator travels *upward* at constant speed, is the tension in the cable greater than, less than, or equal to the weight of the elevator? Give a physics explanation, including both a free-body diagram and reference to appropriate physical principles.

47

b) As an elevator *accelerates upward*, is the tension in the cable greater than, less than, or equal to the weight of the elevator? Explain.

c) As an elevator *accelerates downward*, is the tension in the cable greater than, less than, or equal to the weight of the elevator? Explain.

5.2 Using Newton's Second Law

Shown below are free body diagrams for an object of mass m that has three forces being exerted on it. For each diagram, write the x- and y-components of Newton's Second Law, Eq. 5-2. The force-side of your equations should be written in terms of the *magnitudes* of the forces $|\vec{F}_1|$, $|\vec{F}_2|$, $|\vec{F}_3|$ and any *angles* defined in the diagram. Component a_y is given for 4. to illustrate what you need to do.

4.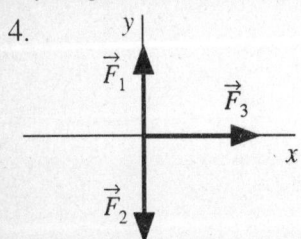

$a_x =$

$a_y = \frac{1}{m}(|\vec{F}_1| - |\vec{F}_2|)$

5.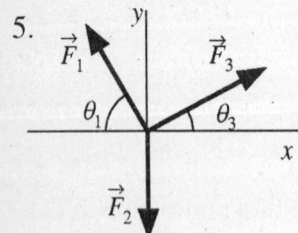

$a_x =$

$a_y =$

6.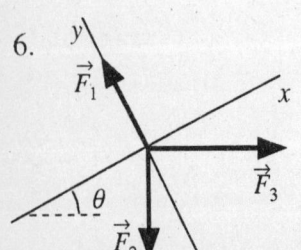

$a_x =$

$a_y =$

Chapter 5 Dynamics I: Newton's Second Law   49

Two or more forces are exerted on a 2 kg object. They are shown below on a free body diagram that includes a grid for accurate measurement of the forces. The units of the grid are newtons. Determine the acceleration $\vec{a}$ of the object for each case. To give your answer:
  a) Draw a vector arrow *on the grid*, starting at the origin, to show the *direction* of $\vec{a}$. (The numerical scale of the grid refers to the forces and is not relevant to $\vec{a}$.)
  b) In the space to the right, determine the numerical values of the component a_x and a_y. Then find the magnitude $|\vec{a}|$.

7.
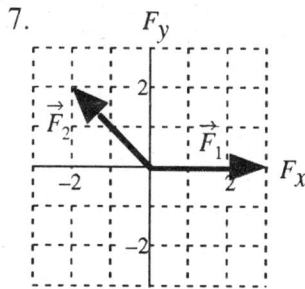

8.
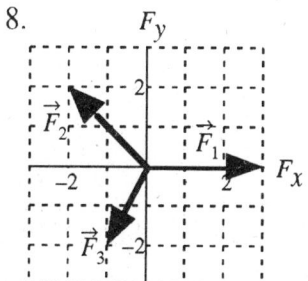

9.
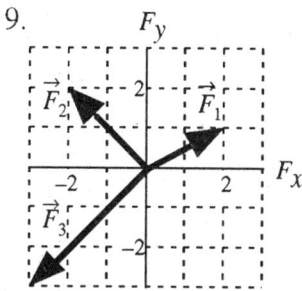

Three forces $\vec{F}_1$, $\vec{F}_2$, and $\vec{F}_3$ are exerted, in the xy-plane, on a 1 kg object. Two of the forces are shown on the free body diagrams below, but the third is missing. The acceleration of the object is specified in each case. For each:
  a) Use a *black* pencil to draw (and label) *on the grid* the missing third force vector.
  b) Use a *red* pencil to draw (and label) *on the grid* the net force vector $\vec{F}_{net}$.

10.
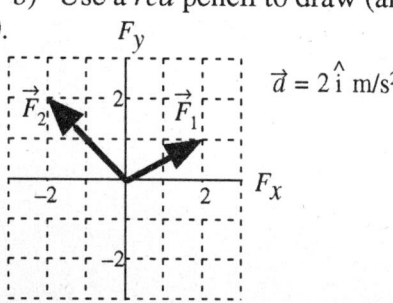

$\vec{a} = 2\hat{i}$ m/s^2

11.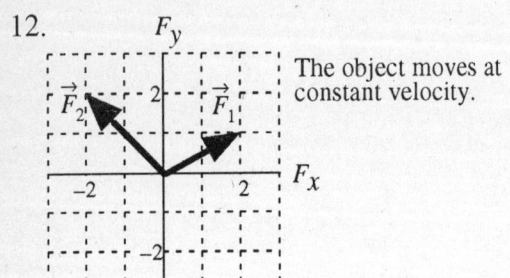

12.

5.3 Mass and Weight

13. Decide whether each of the following are True or False. Give a reason!
a) The mass of an object depends on its location.

b) The weight of an object depends on its location.

c) Mass and weight describe the same thing in different units.

14. a) An astronaut takes his bathroom scales to the moon and then stands on them. Is the reading of the scales his correct weight? Explain.

b) On earth, the "conversion factor" from kilograms to pounds is 1 kg = 2.2 lb. What is the "conversion factor" from kilograms to pounds on the moon, where the acceleration due to gravity on the moon is $g = 1.6$ m/s^2.

15. A 1 kg wood ball and a 10 kg lead ball have identical shapes. They are dropped simultaneously from a tower.
a) As the balls fall, are the forces on them equal in magnitude or different? (Assume that air resistance is negligible.) If different, which has the larger force? Explain.

b) In the absence of air resistance, are their accelerations equal or different? If different, which has the larger acceleration? Explain.

c) Which ball hits the ground first, or are they simultaneous?

d) If air resistance is present, each ball will experience the *same* air resistance force because air resistance depends only on an object's shape — and both have the same shape. Draw free body diagrams for the two balls as they fall in the presence of air resistance. Make sure that your vectors all have the correct *relative* lengths.

e) When air resistance is included, are the accelerations of the balls equal or different? If not, which has the larger acceleration? Explain, using your free body diagrams from d) and Newton's laws.

f) Which ball now hits the ground first, or are they simultaneous? Explain.

16. The terms "vertical" and "horizontal" are frequently used in physics. Give *operational definitions* for these two terms. Recall, from Chapter 1, that an "operational definition" defines a term by how it is measured or determined. Keep in mind that your definition must apply equally well in a laboratory room and on a steep mountainside.

5.4   Equilibrium

17. The vectors below show five forces that can be applied individually or in combinations to an object. Which forces or combinations of forces would cause the object to be in equilibrium?

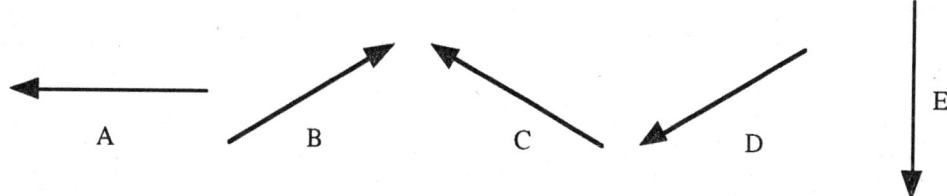

5.5   Apparent Weight

18. As an elevator *descends* it first starts from rest, then moves at a steady speed, then brakes to a halt. During each of these three phases of the motion, is your apparent weight greater than, less than, or equal to your true weight? Explain, using free body diagrams.

19. Suppose you attempt to pour out 100 g of salt, using a pan balance for measurements, while in an elevator that is accelerating upward. Will the quantity of salt be too much, too little, or the correct amount. Explain.

20. Suppose you have a jet-powered flying platform that can move up and down along a vertical line. For each of the following cases, is your apparent weight equal to, greater than, or less than your true weight. Explain.

a) You are ascending and speeding up.

b) You are descending and speeding up.

c) You are ascending at a constant speed.

d) You are descending at constant speed.

e) You are ascending but slowing down.

f) You are descending but slowing down.

21. An box with a 75 kg passenger inside is launched straight up into the air by a giant rubber band. Once the box has left the rubber band but is still moving *upward*:

a) What is the passenger's true weight?

b) What is the passenger's apparent weight?

22. An astronaut orbiting the earth is handed two balls that have identical outward appearances. One, however, is hollow while the other is filled with lead. How might the astronaut determine which is which?

5.6 Friction

23. A block pushed along the floor with velocity $\vec{v}_0$ slides a distance d after the pushing force is removed.

a) If the mass of the block is doubled but the initial velocity is not changed, what is the distance the block slides before stopping?

b) If the initial velocity of the block is doubled but the mass is not changed, what is the distance the block slides before stopping?

24. a) Consider a wheel *rolling* without slipping along a rough surface. Is the friction between the wheel and the surface static friction or kinetic friction? Explain (You will want to think about this one carefully.)

b) To stop a car in the shortest possible distances, it is always recommended that you NOT press the brakes so hard as to lock the wheels and skid. You can stop in a shorter distance if you avoid skidding. (This is why some cars have anti-lock brakes.) Give a physics explanation for this, based on your answer to part a) and what you have learned in this section about friction.

25. a) Consider a crate resting on the floor of a truck. If the truck accelerates slowly, the crate has the same acceleration as the truck. What force or forces are being exerted on the crate to accelerate it? In what direction do those forces point? Draw a free body diagram of the crate.

b) What happens to the crate if the truck accelerates too rapidly? Why does this happen? Give a physics explanation.

5.7   More Examples of the Second Law

No exercises.

22. An astronaut orbiting the earth is handed two balls that have identical outward appearances. One, however, is hollow while the other is filled with lead. How might the astronaut determine which is which?

5.6 Friction

23. A block pushed along the floor with velocity $\vec{v}_0$ slides a distance d after the pushing force is removed.

a) If the mass of the block is doubled but the initial velocity is not changed, what is the distance the block slides before stopping?

b) If the initial velocity of the block is doubled but the mass is not changed, what is the distance the block slides before stopping?

24. a) Consider a wheel *rolling* without slipping along a rough surface. Is the friction between the wheel and the surface static friction or kinetic friction? Explain (You will want to think about this one carefully.)

b) To stop a car in the shortest possible distances, it is always recommended that you NOT press the brakes so hard as to lock the wheels and skid. You can stop in a shorter distance if you avoid skidding. (This is why some cars have anti-lock brakes.) Give a physics explanation for this, based on your answer to part a) and what you have learned in this section about friction.

25. a) Consider a crate resting on the floor of a truck. If the truck accelerates slowly, the crate has the same acceleration as the truck. What force or forces are being exerted on the crate to accelerate it? In what direction do those forces point? Draw a free body diagram of the crate.

b) What happens to the crate if the truck accelerates too rapidly? Why does this happen? Give a physics explanation.

5.7 More Examples of the Second Law

No exercises.

Chapter 6

Dynamics II: Motion in a Plane

6.1 Kinematics in Two Dimensions

1. Complete the motion diagram for this trajectory, showing velocity and acceleration vectors.

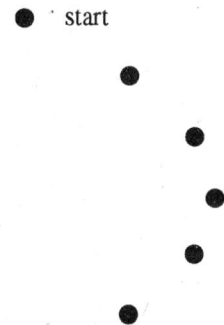

2. Shown below are the x-versus-t graph and the y-versus-t graph for a particle moving along a trajectory in the xy-plane.

a) Use the grid below to draw a y-versus-x graph of the trajectory.

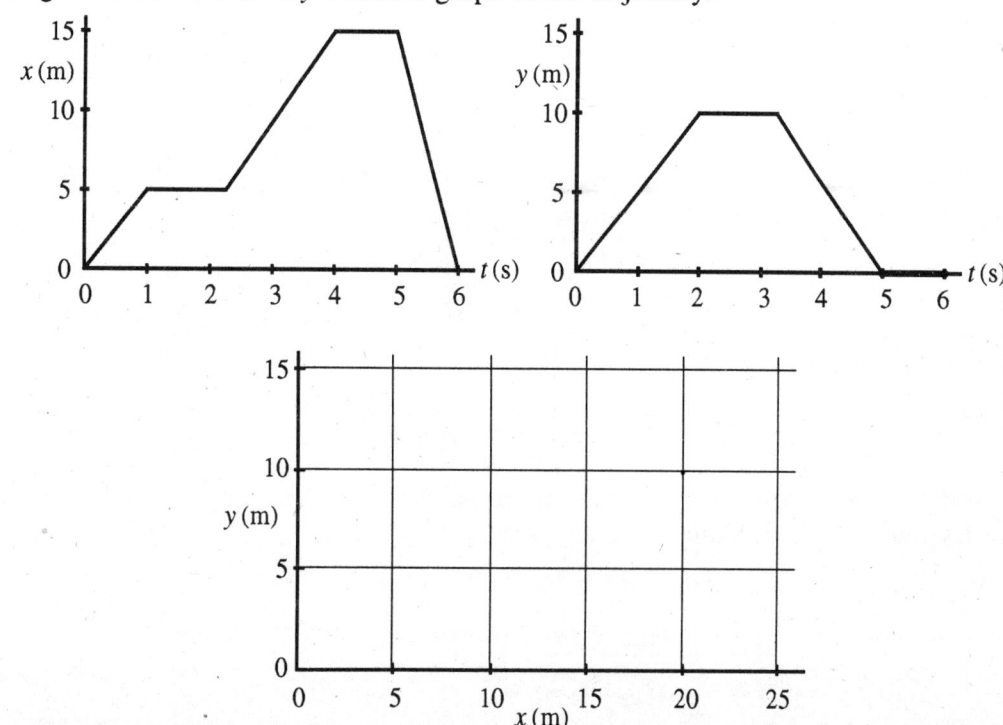

58   Chapter 6 Dynamics II: Motion in a Plane

b) Draw the velocity vector at $t = 3.5$ s on your graph.
c) What is the particle's speed at $t = 3.5$ s?

d) What is the particle's acceleration at $t = 3.5$ s?

3. The trajectory of a particle is shown below. The particle's position at 1 second intervals is indicated with dots. The particle moves between each pair of dots at constant speed.

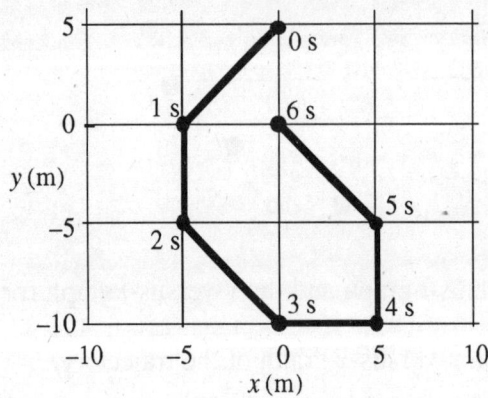

a) Use the two sets of axes below to draw x-versus-t and y-versus-t graphs for the particle.

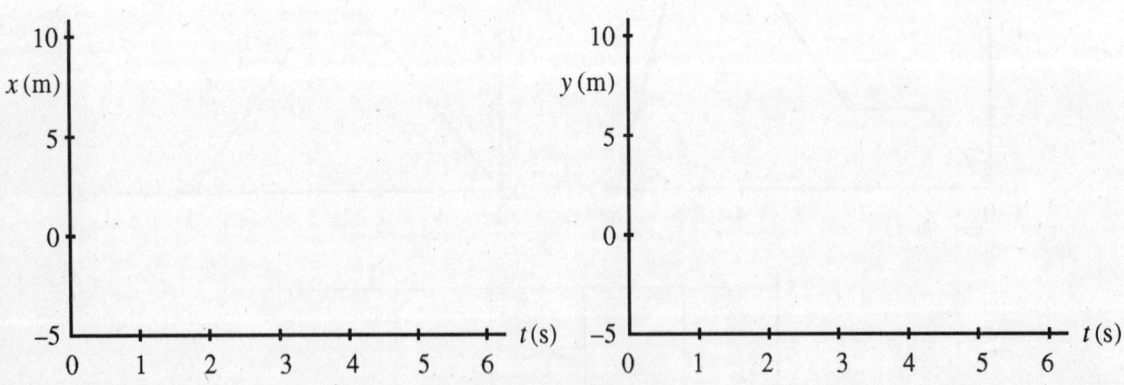

b) Is the particle's speed between $t = 5$ s and $t = 6$ s greater than, less then, or equal to its speed between $t = 1$ s and $t = 2$ s? Explain

4. The figure shows a ramp and a ball that rolls along the ramp. Draw vector arrows on the figure to show the ball's acceleration at each of the lettered points A to E (or write $\vec{a} = 0$, if appropriate).

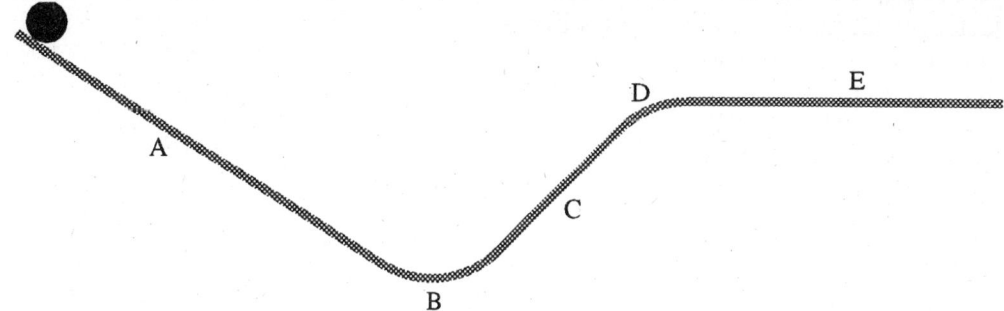

5. If you know the position vectors of a particle at two points on its trajectory and you also know the time it took to get from the first point to the second, can you determine the particle's average velocity? Its instantaneous velocity? Its acceleration? Explain.

6.2 Dynamics in Two Dimensions

6. A rocket motor is taped to a hockey puck. The rocket is oriented so that its thrust is to the left. The puck is given a push across frictionless ice as shown in the bird's-eye view. The rocket will be turned on by remote control as the puck crosses dotted line #2, then turned off as it crosses dotted line #3. Sketch the puck's trajectory from line #1 until it crosses line #4.

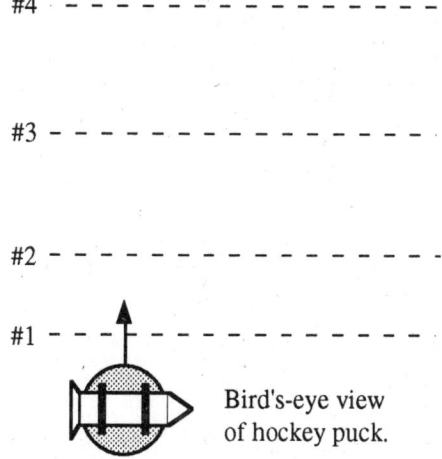

7. The same puck, but now without the rocket, is again pushed across the ice in the same direction. The puck receives a sharp, very short kick toward the right as it crosses line #2. It receives a second kick, of equal strength and duration but directed toward the left, as it crosses line #3. Sketch the puck's trajectory from line #1 until it crosses line #4.

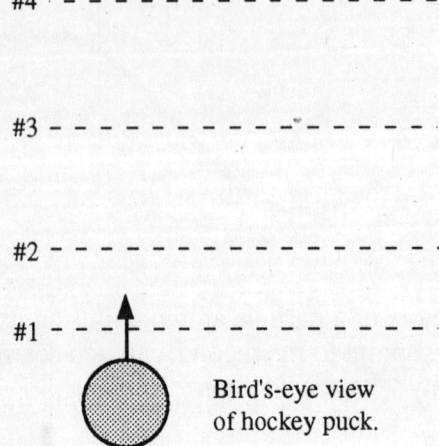

8. Tarzan swings through the jungle while hanging from a vine.

a) Use a motion diagram, as in Chapter 1, to find the direction of Tarzan's acceleration vector $\vec{a}$
   i) Immediately after stepping off the branch, and
   ii) At the exact bottom of his swing.

b) At each of these two points, is the magnitude of the tension in the vine $|\vec{T}|$ greater than, less than, or equal to Tarzan's weight? Give a physics explanation, using Newton's laws. (Hint: Use a coordinate system at each point where $\vec{a}$ points along one axis.)

6.3 Projectile Motion

9. As a projectile moves along a parabolic trajectory

a) Is there any point on the trajectory where $\vec{v}$ and $\vec{a}$ are parallel to each other? If so, where?

b) Is there any point where $\vec{v}$ and $\vec{a}$ are perpendicular to each other? If so, where?

c) Which of the following remain constant throughout the entire trajectory: $|\vec{r}|, x, y, |\vec{v}|, v_x, v_y, \vec{a}$?

10. The figure shows a ball that rolls down a quarter-circle ramp, then off a cliff. Sketch the ball's trajectory from when it is released until it hits the ground.

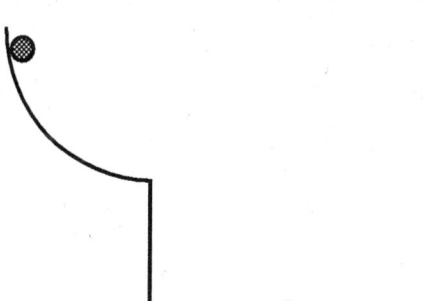

11. a) A cart that is rolling along at constant velocity fires a ball straight up. When the ball returns, will it land in front of the launching tube, behind the launching tube, or directly in it? Give a physics explanation.

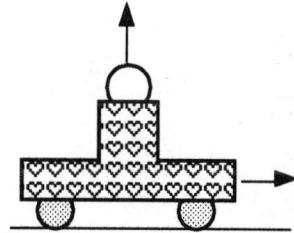

b) If the cart is accelerating in the forward direction, will your answer change? If so, how?

12. Five balls are released simultaneously from the same height h above the ground. Balls 1 - 4 all have the same initial *speed* but are launched at, respectively, angles of 45°, 0°, –45°, and –90°. Ball 5 is released from rest. What is the *order*, from first to last, in which they hit the ground? (Some may be simultaneous.)

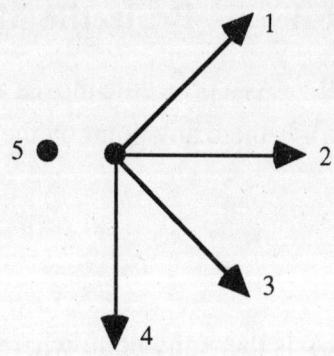

13. A rock is thrown downward, from a bridge, at 30°.

a) Sketch the trajectory on the figure.

b) Immediately after the rock is released, is the magnitude of its acceleration greater than, less than, or equal to g? Explain.

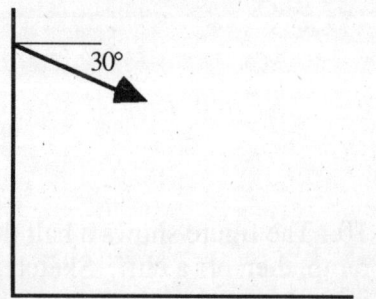

c) At the instant of impact is the rock's speed greater than, less than, or equal to the speed with which it was released? Explain.

6.4　Uniform Circular Motion

14. Find the frequencies, in revolutions per second, of these circular motions:

a) The moon orbiting the earth ($T = 27.3$ days).

b) The earth rotating on its axis.

c) The earth orbiting the sun.

15. a) The crankshaft in your car rotates at 3000 rpm. What is the frequency in revolutions per second?

b) A record turntable rotates at 33.3 rpm. What is the period in seconds?

16. A ball on a string moves in a vertical circle.

a) When the ball is at the very top, is the tension in the string greater than, less than, or equal to the ball's weight? Give a physics explanation.

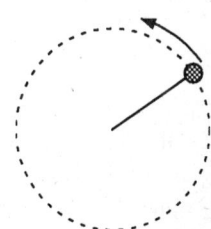

b) When the ball is at the very bottom, is the tension in the string greater than, less than, or equal to the ball's weight? Give a physics explanation.

17. a) The figure on the left shows a *top view* of a plastic tube that is fixed on a *horizontal* table top. A marble is shot into the tube at A with high speed. Sketch the marble's trajectory after it passes B.

b) The figure on the right shows a ball that is swung in a *vertical* circle on a string. During one revolution, a very sharp knife is used to cut the string at the instant it is hanging vertically down. Sketch the subsequent trajectory of the ball until it hits the ground.

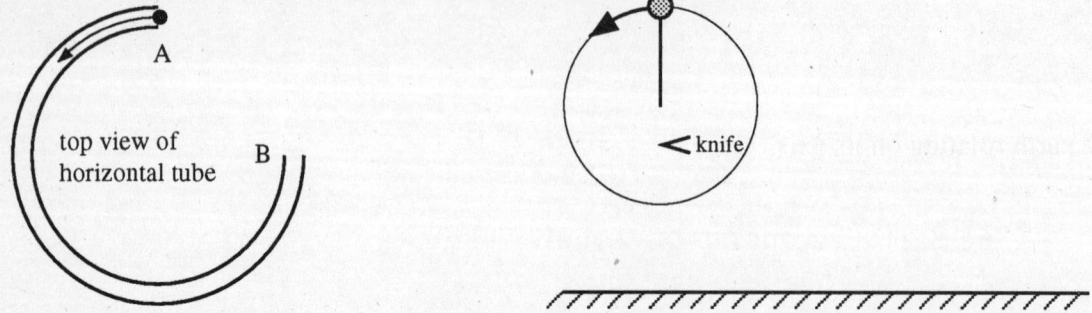

18. A jet airplane is flying on a level course at *constant* velocity.

a) What is the *net* force on the plane? Why?

b) Draw a picture, and identify all of the forces acting on the plane and their directions.

c) Draw a free body diagram. Is your diagram consistent with your answer to a)?

d) Airplanes "bank" when they turn. Why do they do this? Give a physics explanation in terms of forces and physical laws. (Hint: What would a free body diagram look like to an observer *behind* the plane?)

19. Why does mud fly off of a rapidly-spinning car tire?

6.5 Circular Motion and Apparent Weight

20. A stunt plane does a vertical loop-the-loop. At which point in the circle does the pilot feel the heaviest? Give a physics explanation, including a free body diagram.

21. a) You can whirl a bucket of water in a *vertical* circle, passing over your head, without spilling any as long as you whirl it fast enough. What keeps the water in the bucket? Give as complete an explanation as you can, making reference to forces and physical laws.

b) As you slow the bucket down, there comes a speed at which the water just begins to spill out of the bucket at the top of the circle. Describe the "threshold condition" at which this happens.

6.6 Orbits

22. A small projectile is launched parallel to the ground at height $h = 1$ m with sufficient speed to orbit a completely smooth, airless planet. A bug rides in a small hole inside the projectile. Is the bug weightless? Give a physics explanation.

Chapter 7

Dynamics III: Newton's Third Law

7.1 Interacting Systems

7.2 Identifying Interaction Forces

For Exercises 1 - 5:
  a) Draw a picture showing each relevant object *separate* from all other objects, but in the correct spatial orientation. Include the earth and, if appropriate, any surfaces.
  b) Identify *all* forces and show them as *black* or *blue* vectors on the objects. Label each vector as the appropriate $\vec{F}_{\text{A on B}}$.
  c) Connect all action/reaction pairs with *red* dotted lines.

Note: Your pictures should look similar Figs. 7-2 and 7-3.

1. a) A bat is hitting a ball. (Draw your picture from the perspective of someone seeing the *end* of the bat at the moment it strikes the ball.)

   b) The ball then sails through the air.

2. a) A ball is held in your hand,

b) then is released to fall,

c) then bounces as it hits the ground. (Consider the instant of contact with the ground.)

3. A boy pulls a wagon by a rope attached to the front of the wagon. The rope is parallel to the ground. Friction is not negligible.

4. A bicycle accelerates forward from rest. (Treat the bicycle and its rider as a single object.)

5. a) A crate is in the back of a truck as the truck accelerates forward. (Treat the crate and the truck as separate systems.)

b) Identify the specific force that is responsible for accelerating the crate in the forward direction?

6. You are in the middle of a frozen lake with a surface so slippery ($\mu_s = \mu_k = 0$) that you cannot walk. It is nearly dark. You happen to have several rocks in your pocket. The ice is extremely hard, so it cannot be chipped, and the rocks slip on it just as much as your feet do. Can you think of a way to get to shore? Use pictures, forces, and Newton's laws to explain your reasoning.

7. How do you paddle a canoe in the forward direction? Give a physics explanation, including pictures showing forces on the water and forces on the paddle.

8. When you blow up a balloon and release it, it shoots forward. Why? Give a physics explanation, including pictures showing forces on the balloon and forces on the parcel of air that was just released from the balloon.

9. How does a rocket take off? What is the upward force on it? Give a physics explanation, including pictures showing forces on the rocket and forces on the parcel of hot gas that was just released from the rocket.

10. How do basketball players jump straight up in the air? As in the last few questions, give a physics explanation in terms of forces and of action/reaction pairs.

7.3 Newton's Third Law

11. Block A is pushed across a horizontal surface at a *constant* speed by a hand, which exerts force $\vec{F}_{\text{H on A}}$. The surface has friction.

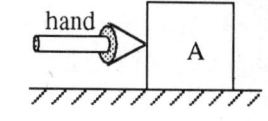

a) Draw two free body diagrams, one for the hand and the other for the block. On these diagrams:
    i) Show *only* the *horizontal* forces, such as was done in Fig. 7-11 of the text.
    ii) Label force vectors using the form $\vec{F}_{\text{C on D}}$.
    iii) Connect action/reaction pairs with dotted lines.
    iv) Make sure vector lengths correctly describe the relative magnitudes of the forces.

b) Rank order *all* of the horizontal forces you showed in a) on the basis of their magnitudes, from the largest to the smallest. For example, if $|\vec{F}_{\text{C on D}}|$ is the largest of three forces while $|\vec{F}_{\text{D on C}}|$ and $|\vec{F}_{\text{D on E}}|$ are smaller but equal, you can record this as $F_{\text{C on D}} > F_{\text{D on C}} = F_{\text{D on E}}$. Explain your reasoning.

c) Now repeat both part a) and part b) for the case that the block is speeding up.

12. A second block B is placed in front of Block A of question 11. B is more massive than A, with $m_B > m_A$. The blocks are speeding up.

a) Consider a *frictionless* surface. Draw three *separate* free body diagrams for the hand, A, and B. Show only the horizontal forces. Label forces in the form $\vec{F}_{C \text{ on } D}$. Use dotted lines to connect action/reaction pairs.

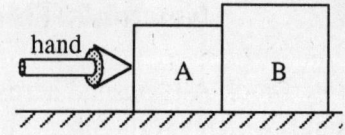

b) By applying the second law to each block and the third law to each action/reaction pair, rank order *all* of the horizontal forces, from largest to smallest. Explain your reasoning. (Hint: A and B have different masses but equal accelerations.)

c) Repeat a) and b) for the case that the surface has friction.

13. Blocks A and B are held on the palm of your outstretched hand and your hand is lifting them vertically at *constant speed*. Assume $m_B > m_A$.

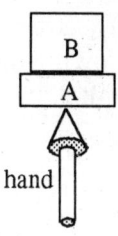

a) Draw three *separate* free body diagrams for your hand, A, and B. Show *all* vertical forces on the block's, including their weights. For your hand, show only forces exerted by the blocks; neglect the weight of your hand or any forces exerted on your hand by your arm. Make sure vector lengths are appropriate. Label forces in the form $\vec{F}_{\text{C on D}}$. Connect action/reaction pairs with dotted lines.

b) Rank order *all* of the vertical forces, from the largest to the smallest. Explain your reasoning.

14. A red car and a blue car, traveling with equal speeds, collide head-on and come to rest. How does the force the red car exerts on the blue car compare to the force that the blue car exerts on the red car? Are they equal, or is one larger than the other? Explain your reasoning.

15. A mosquito collides head-on with a car traveling 60 mph.

a) How do you think the force that the car exerts on the mosquito compares to the force that the mosquito exerts on the car? Why?

74   Chapter 7 Dynamics III: Newton's Third Law

b) Draw *separate* free body diagrams of the car and the mosquito at the moment of collision, showing only horizontal forces. Label forces in the form $\vec{F}_{\text{C on D}}$. Connect action/reaction pairs with dotted lines.

c) Does your answer to b) confirm your answer to a)? If so, explain how. If not, revise your answer to either a) or b) in the space below.

Write the acceleration constraint, in terms of *components*, for each of the situations in Exercises 16 - 20. That is, write $(a_1)_x = (a_2)_x$, if that is the appropriate answer, rather than $\vec{a}_1 = \vec{a}_2$.

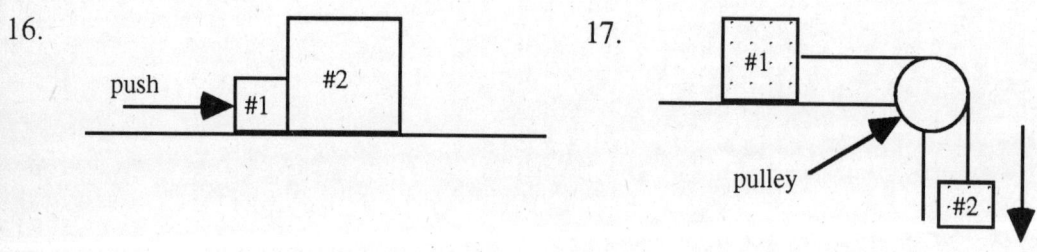

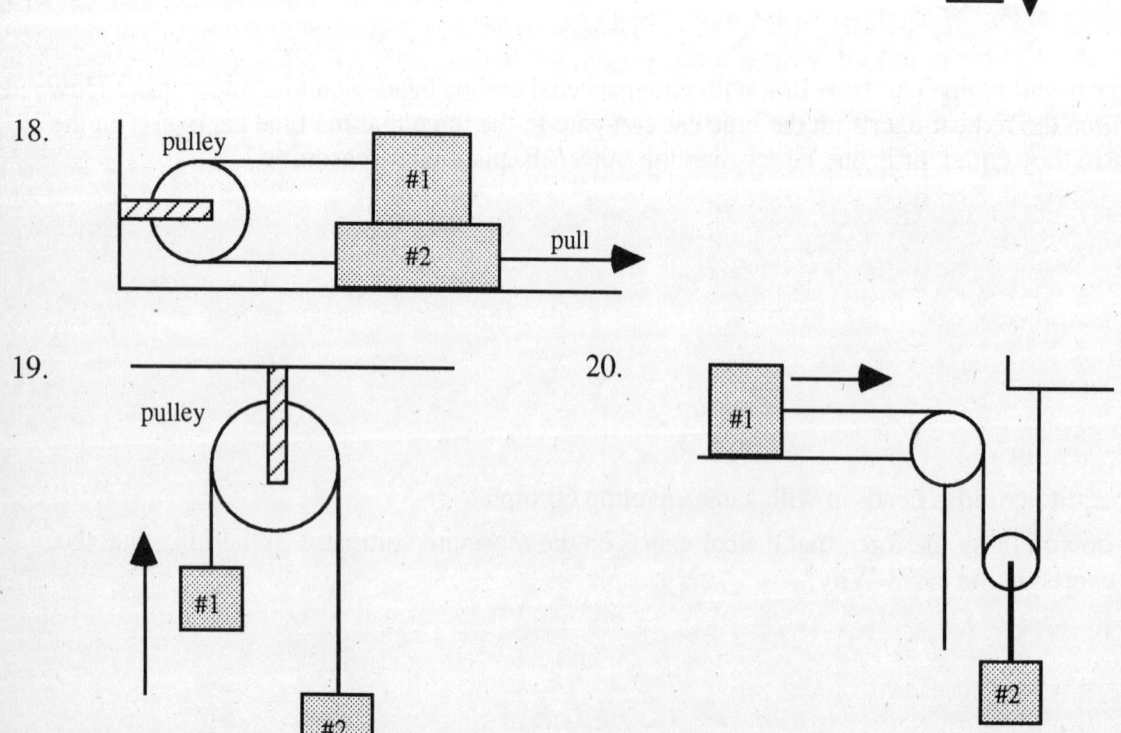

7.4 Interactions with Strings and Ropes

For Exercises 21 - 26, all the masses are at rest, the strings and pulleys are massless, and the pulleys are frictionless. Determine the reading of the spring scale in each exercise.

21.
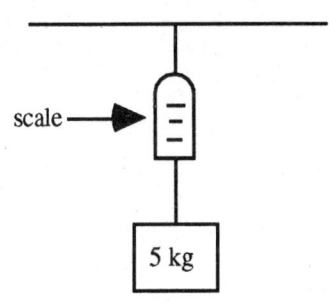

22.
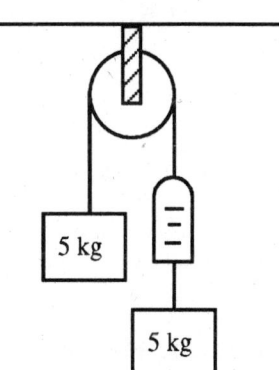

23.
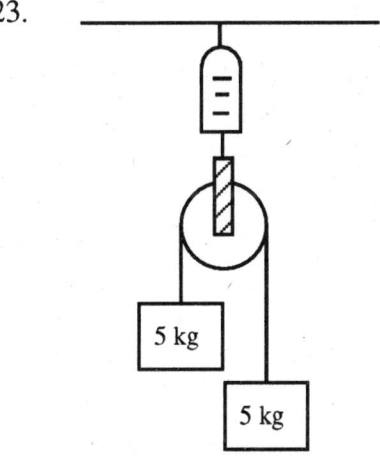

24.
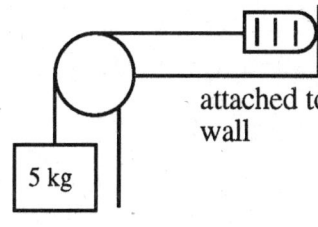

attached to wall

25.
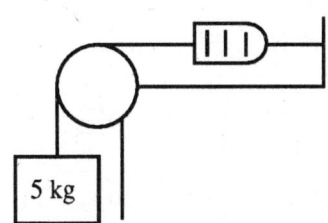

26.
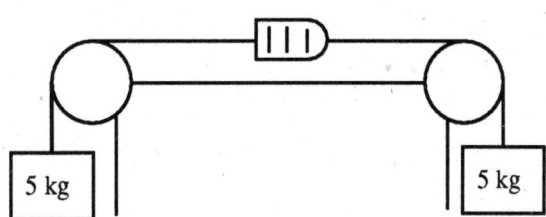

27. a) A tight-rope walker at the circus steps onto the high wire, causing it to sag slightly. Is the tension in the wire less than, greater than, or equal to the performer's weight? Give a physics explanation. Include a free body diagram.

b) The leading circus magazine advertises a new wire made of a material called DreamRope. The ad says that a DreamRope wire will remain perfectly straight and horizontal, with absolutely no sag, as the performer walks across. Should you order some? Explain.

7.5 A Revised Strategy

No exercises

7.6 Examples of Third Law Problems

28. Blocks A and B, with $m_B > m_A$, are connected by a string. A hand pushing on the back of A accelerates them along a frictionless surface. The string (S) is *not* massless.

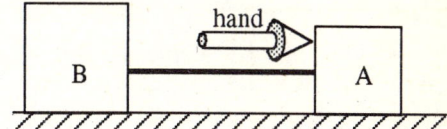

a) Draw separate free body diagrams for A, S, and B, showing only horizontal forces. Be sure vector lengths are appropriate. Connect any action/reaction pairs with dotted lines.

b) Use Newton's second and third laws to rank order all of the horizontal forces from the largest to the smallest. Explain your reasoning.

c) Repeat a) and b) if the string is now massless.

d) You might expect to find $F_{S \, on \, B} > F_{H \, on \, A}$ because $m_B > m_A$. Did you find this in b) and c)? Explain why $F_{S \, on \, B} > F_{H \, on \, A}$ is or is not a correct statement.

29. Blocks A and B are connected by a massless string over a massless, frictionless pulley. The blocks have just this instant been released from rest.

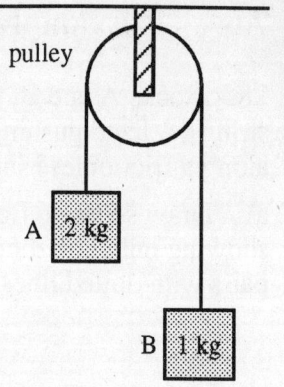

a) Will the blocks accelerate? If so, in which directions?

b) Draw a separate free body diagram for each block. Make vectors the appropriate lengths and connect any action/reaction pairs or "as if" action/reaction pairs with dotted lines.

c) Rank order all of the vertical forces. Explain your reasoning.

d) Compare the magnitude of the *net* force on A with the *net* force on B. Are they equal, or is one larger than the other? Explain.

e) Is the acceleration of the block that falls less than, greater than, or equal to *g*? Explain.

30. A man accelerates a crate, which has mass $m_C < m_M$, across the floor with a *massless* rope. The floor surface does have friction — since otherwise the man could not walk! Assume the rope is parallel to the ground.

a) Draw separate free body diagrams of the man and the crate, showing only *horizontal* forces. Connect any action/reaction pairs or "as if" action/reaction pairs with dotted lines.

b) How does the magnitude of $\vec{F}_{\text{M on C}}$ compare to that of $\vec{F}_{\text{C on M}}$? Explain.

c) Is there a *net* force on the man? If so, in which direction?

d) Is there a *net* force on the crate? If so, in which direction?

e) If you answered Yes in c) and d), which net force is larger? Why?

31. A very smart three-year-old child is given a wagon for her birthday. She refuses to use it. "After all," she says, "Newton's third law says that no matter how hard I pull, the wagon will exert an equal but opposite force. So I will never be able to get it to move forward." What would you say to her in reply?

32. Will hanging a magnet in front of an iron cart make it go? Explain why or why not.

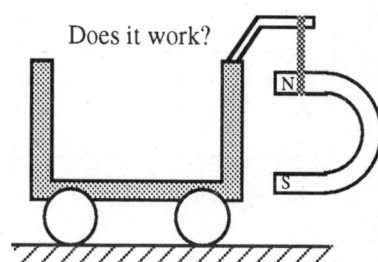

Does it work?

33. In case a, Block A is accelerated across a frictionless table by a hanging 10 N weight. In case b, the same Block A is accelerated by a steady 10 N tension in the string.

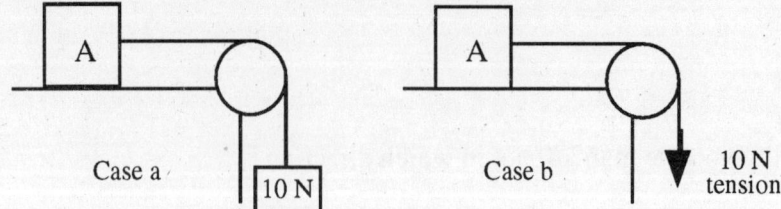

Is Block A's acceleration in case b greater than, less than, or equal to its acceleration in case a? Explain.

For Exercises 34 - 35, draw separate free body diagrams for objects #1 and #2. Connect any action/reaction pairs (or forces that act "as if" they are action/reaction pairs) together with dotted lines.

34.

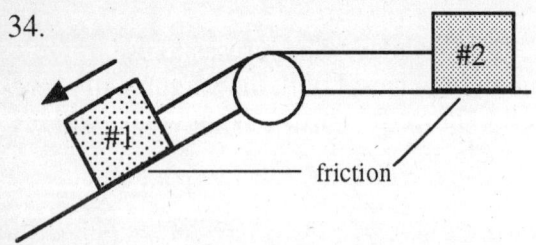

35.

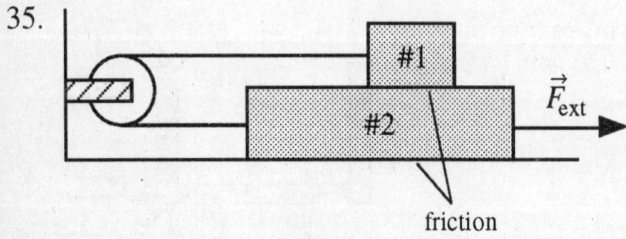

Chapter 8

Momentum And Its Conservation

8.1 Conservation Laws

8.2 Impulse and Momentum

1. Explain the concept of *impulse* in non-mathematical language. That is, do not simply put an equation in words and say that "impulse is the time-integral of force." Explain it in terms of *physical* concepts such as you might to an educated person who had never heard of it.

2. A 2 kg object is moving to the right with a speed of 1 m/s when it suddenly experiences an impulse. What is the object's speed and direction after the impulse? Answer this question for the impulses shown in a) - d).

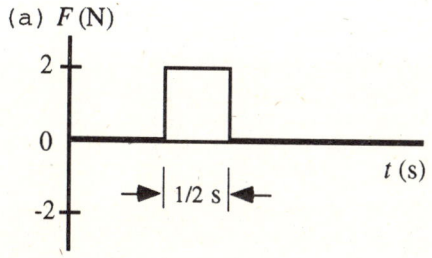
(a)

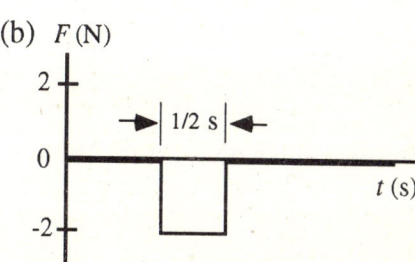
(b)

(c)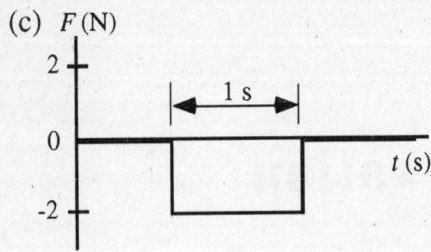

(d)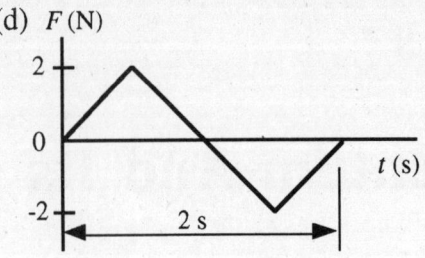

3. Describe what happens in each of the following situations.
   i) First describe it using the language of force-acceleration-action/reaction.
   ii) Then describe it in the language of impulse-momentum.
Use both words and pictures in your explanation.

a) A blob of clay is thrown at a stationary bowling ball.

b) A falling rubber ball bounces off the floor.

c) Two equal masses are pushed apart by a compressed spring between them.

8.3 Momentum in Two Particle Collisions

4. A small, light ball S and a large, heavy ball L move toward each other, as shown, and undergo an elastic collision.

a) How does the force that S exerts on L compare to the force that L exerts on S? That is, is $F_{S\,on\,L}$ larger, smaller, or equal to $F_{L\,on\,S}$? Give a physics explanation.

b) How does the time interval during which S experiences a force compare to the time interval during which L experiences a force?

c) Sketch a graph showing a *plausible* $F_{L\,on\,S}$ as a function of time and another graph showing a plausible $F_{S\,on\,L}$ as a function of time. Be sure think about the *sign* of each force.

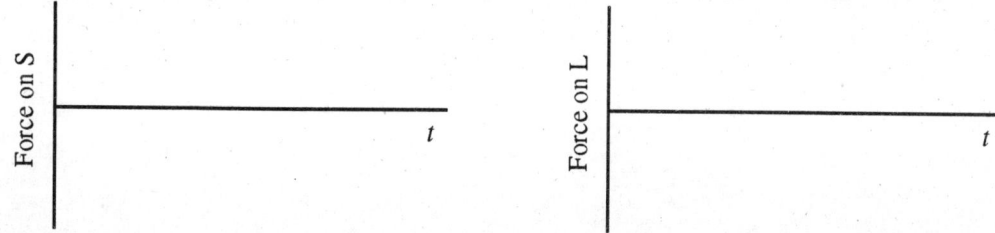

d) How does the impulse delivered to S compare to the impulse delivered to L? Explain.

e) How does the momentum change of S compare to the momentum change of L? Explain.

f) How does the velocity change of S compare to the velocity change of L?

g) What is the change in the total momentum of the system — positive, negative, or zero?

5. Two particles collide, one of which was initially at rest.
a) Is it possible for both particles to be at rest after the collision? Give an example in which this happens or give a physics explanation as to why it can't happen.

b) Is it possible for one particle to be at rest after the collision? Give an example in which this happens or give a physics explanation as to why it can't happen.

8.4 Conservation of Momentum

6. Explain the concept of "isolated system" in non-mathematical language.

7. A golf club continues forward after hitting the ball. Is momentum be conserved in the collision? Explain, making sure you are careful to identify the "system."

8. As you release a ball, it falls — gaining speed and momentum. Is momentum conserved?
a) Answer this question from the perspective of choosing the ball alone as "the system."

b) Answer this question from the perspective of choosing ball+earth as "the system."

9. Just before the ball of Exercise 8 bounces, its momentum is directed downward. Just after the bounce the ball's momentum is directed upward. If the bounce is a collision, why does it appear that momentum is not being conserved?

Prepare a Pictorial Model for Exercises 10 - 12, but do *not* solve them. Your Pictorial Models should include sketches of "before" and "after," should define symbols relevant to the problem, should list known information, and should identify the desired unknown. If this is a "two-part" problem, with a subsequent dynamics problem, your Pictorial Model should include all relevant information for both parts.

10. A 50 kg archer, standing on frictionless ice, shoots a 100 g arrow at a speed of 100 m/s. What is the recoil velocity of the archer?

11. The parking brake on a 2500 lb Cadillac has failed, and it is rolling slowly, at 1 mph, toward a group of small innocent children. As you see the situation, you realize there is just time for you to drive your 1000 lb Volkswagon head-on into the Cadillac and thus to save the children. With what speed should you impact the Cadillac to just bring it to a halt?

12. Fred Fingers, 60 kg, is running upfield with the football at a speed of 6 m/s when he is met head-on by Brutus, 120 kg, who is moving with a speed of 4 m/s. Brutus grabs Fred in a tight grip, and they fall to the ground. Which way do they slide, and how far? The coefficient of kinetic friction between football uniforms and Astroturf is 0.3.

8.5 Recoil, Radioactivity, and Rockets

8.6 Collisions in Two Dimensions

8.7 Rutherford and the Structure of Atoms

13. The text claims that the nucleus takes up only about one part in 10^{11} of the volume of an atom. Demonstrate that this is true *without* calculating either the atom's volume or the nucleus' volume. The radius of a nucleus is $\approx 10^{-14}$ m while the radius of an atom is 5×10^{-11} m.

14. Write a brief summary of *how we know* that atoms have a very small, heavy positive nucleus surrounded by light, negative electrons.

Chapter 9

Work and Energy

9.1 Introduction

1. One month Jose has income of $3000, expenses of $2500, and he sells $300 of stocks.

a) Can you determine Jose's liquid assets L at the end of the month? If so, what is L? If not, why not?

b) Can you determine the amount by which Jose's liquid assets *changed* during the month? If so, what is ΔL?

2. Jose begins the month with $2000 of liquid assets and $5000 of savings. His financial activity for the month looks as follows:

Day of Month	Activity
1	Receives $3000 paycheck; deposits in checking
3	Spends $500
8	Buys a $1000 savings bond
10	Pays bills totaling $1000
15	Receives $100 birthday present from Grandma
23	Sells $1500 of stock
28	Buys a $1200 bicycle

a) What are Jose's liquid assets and savings at the end of the month?

b) Show that Jose's law of conservation of wealth is satisfied.

9.2 Energy

9.3 The Basic Energy Model

3. Upon what basic quantity does kinetic energy depend? _____

4. Upon what basic quantity does potential energy depend? _____

5. What are the two primary processes by which energy can be transferred from the environment to a system?

6. What is meant when we say that the law of conservation of energy is a scientific "hypothesis?"

7. What is meant by an *isolated system*?

8. a) During one month, Jose transfers money out of savings *and* has an income that exceeds his expenditures. Can you conclude anything about the change of his liquid assets during this month? That is, can you determine if L increases, decreases, or stays the same? Explain.

b) A process occurs in which the potential energy *decreases* while work is done by the environment *on* the system. Can you conclude anything about the change of kinetic energy? That is, can you determine if the kinetic energy increases, decreases, or stays the same? Explain.

9. A process occurs in which the potential energy *increases* while work is done by the environment on the system. Can you conclude anything about the change of kinetic energy?

10. The kinetic energy of a system decreases while its potential energy does not change. What is doing work on what? That is, does the environment do work on the system or does the system do work on the environment? Explain.

9.4   Kinetic Energy

11. Can kinetic energy ever be negative? _____
Give a plausible *reason* for your answer without making use of any formulas.

12. a) If a particle's velocity suddenly increases by a factor of three, by what factor does its kinetic energy change?

b) Particle A has half the mass and eight times the kinetic energy of particle B. How do the velocities of A and B compare?

13. On the axes below, draw graphs the kinetic energy of
a) A 1000 kg car that uniformly accelerates from 0 to 20 m/s in 20 s.
b) A 1000 kg car moving at 20 m/s that brakes to a halt with uniform deceleration in 4 s.
c) A 1000 kg car that drives once around a 40 m diameter circle at a speed of 20 m/s.
Calculate K at several times, plot the points, and draw a smooth curve between them.

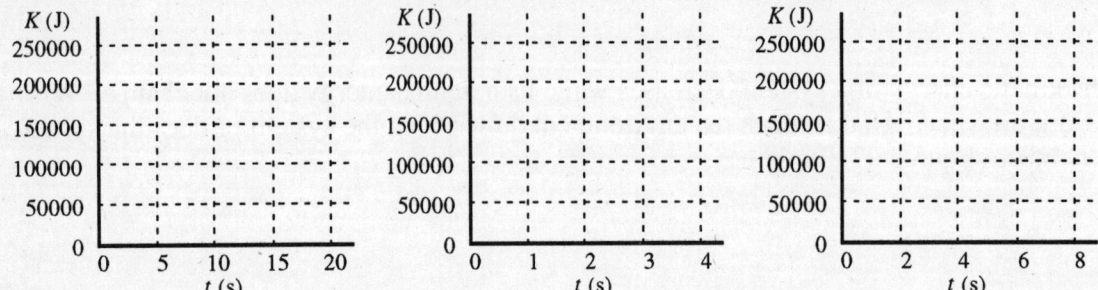

9.5   Work

14. For each of the situations described below:
     i) Draw a diagram, similar to text Figs. 9-4 or 9-5,
     ii) Identify *all* forces acting on the particle, and
     iii) Determine if the work done by each of these forces is positive (+), negative (–), or zero (0). Make a table beside the figure showing *each* force and the sign of its work.

a) An elevator moves upward.

b) An elevator moves downward.

c) You push a box across a rough floor.

d) You slide down a steep hill.

e) A ball is thrown straight up. Consider the ball from one microsecond after it leaves your hand until the top point of its trajectory.

f) A ball is thrown straight up. Consider the ball from the top of its trajectory until one microsecond before you catch it on the return.

g) A car turns a corner at constant speed.

9.6 The Vector Dot Product

15. For each of the pairs of vectors shown, determine if $\vec{A} \cdot \vec{B}$ is +, –, or 0.

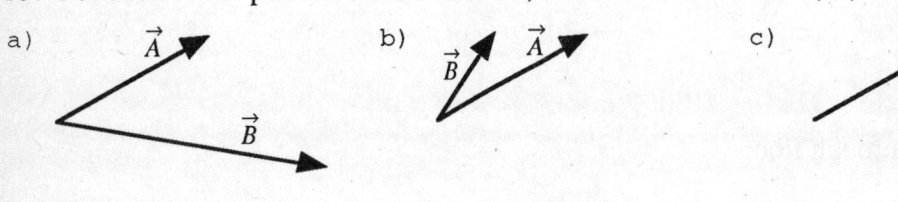

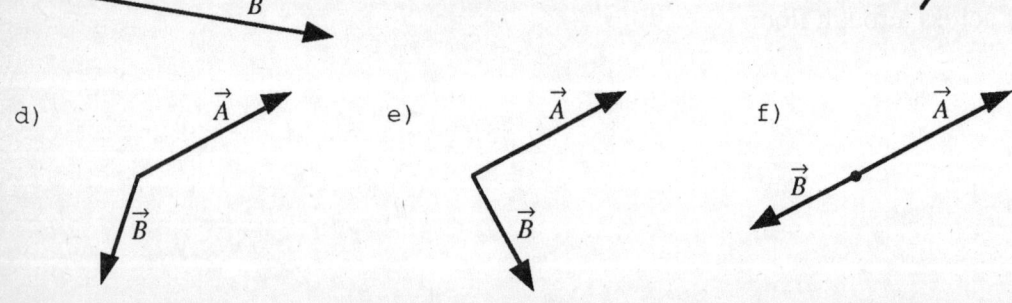

16. Each of the diagrams below shows a vector $\vec{A}$. Draw a vector $\vec{B}$ that will cause $\vec{A} \cdot \vec{B}$ to have the sign indicated.

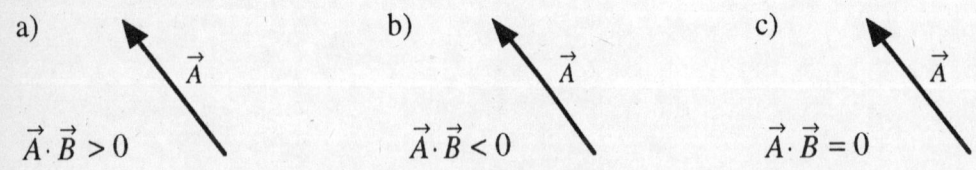

9.7　The Work-Kinetic Energy Theorem

17. For each of the situations described:

   a) Draw a pictorial model, similar to text Figs. 9-12 – 9-14, showing the object at the beginning (before) and end (after) of its motion. Label "before" and "after."
   b) Show and label the displacement Δx with an arrow on the diagram.
   c) Draw a free body diagram showing all forces on the object.
   d) Make a table to show the sign (+, –, or 0) of
   　　i) W for each force on the free body diagram,
   　　ii) W_{net}, and
   　　iii) ΔK.

a) A ball rolls to a stop along a horizontal floor with friction.

b) A ball rolls down a frictionless slope.

c) A ball rolls up a frictionless slope.

d) A ball falling after being released from rest.

e) A ball rising after being tossed straight up.

f) A descending elevator braking to a halt.

g) A rocket being launched straight up.

18. A 0.5 kg mass on a 1 meter long string is swinging in a circle on a horizontal frictionless table at a speed of 2 m/s.
a) How much work is done on the mass by the tension in the string during one revolution? Explain.

b) Is your answer to a) consistent with the work-kinetic energy theorem?

9.8 Work Done By a Continuously-Varying Force

19. In Chapter 8, we found a graphical interpretation of Δp as the area under the F-versus-t graph from an initial time t_i to a final time t_f. Can you provide an analogous graphical interpretation of ΔK, the change in kinetic energy? Explain your reasoning.

20. A 1 kg particle moving along the x-axis with an initial velocity of 2 m/s at $x = 0$ is subjected to the force shown in the graph. What is the particle's velocity when it gets to $x = 5$ m?

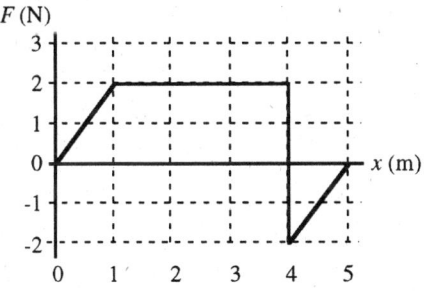

21. Particle A, which has less mass, and particle B, which has more mass, are each pushed by equal forces through a distance of 1 m. Both start from rest.

a) Compare the amount of work done on each particle. (Here, and in subsequent questions, "compare" means to say which is larger and which smaller or, if appropriate, to say that the two values are equal.)

b) Compare the impulses delivered to each particle.

c) Compare their final velocities.

d) Compare the time it takes each particle to cover the 1 m distance.

22. Particle A, which has less mass, and particle B, which has more mass, are each pushed by equal forces for a time of 1 s. Both start from rest.

a) Compare the amount of work done on each particle.

b) Compare the impulses delivered to each particle.

c) Compare their final velocities.

d) Compare the distance traveled by each during the 1 s interval..

9.9 Restoring Forces and Hooke's Law

23. An elastic object is attached to the floor and pulled straight up with a string. The string's tension is measured. A graph of the tension data is shown.

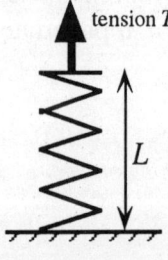
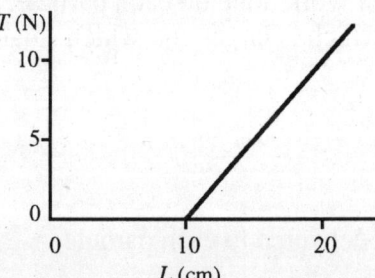

a) Does this elastic object obey Hooke's Law? Explain why or why not.

b) If it does, what is the spring constant?

c) What is L_0, the spring's unstretched length?

24. Draw a figure analogous to text Fig. 9-19 for a spring that is fixed on the *right* end. Use the figure to demonstrate that F and Δs have opposite signs for both stretching and compression.

25. A spring has an unstretched length of 10 cm. It exerts a restoring force F when stretched to a length of 11 cm.
a) At what length is the spring's restoring force $3F$?

b) At what compressed length is the restoring force $2F$?

26. Bob applies a 200 N force to a spring whose left end is fixed in position. His pull stretches the spring 20 cm. The same spring is then used for a tug-of-war between Bob and Bill. They each pull on their end of the spring with a 200 N force.
a) How much does Bob's end of the spring move?

b) How much does Bill's end of the spring move?

c) Is your answer consistent with Hooke's law? Explain.

27. In Example 9-9 in the text, a compressed spring having a spring constant 10 N/m expands from $x_1 = -10$ cm $= -0.10$ m to its equilibrium position at $x_2 = 0$.

a) Graph $F_{\text{spring on ball}}$ from $x_1 = -0.10$ m to $x_2 = 0$.

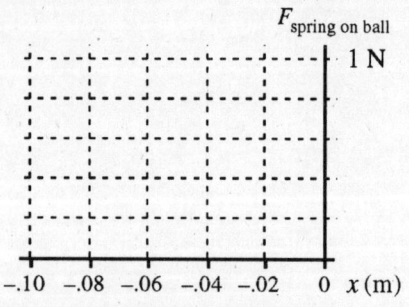

b) Use your graph to determine ΔK, the change in kinetic energy, for a ball launched by the spring after the spring is compressed by 10 cm.

c) Use your result to b) to find the launch speed for a 10 g ball. Compare your answer to the value found in the Example 9-9.

28. In Example 9-9 in the text, the work done by the spring on the ball was found to be $W = (1/2)kx_1^2$ where x_1 was the initial position of the ball. Because x_1 is squared, we can write this as simply $W = (1/2)kd^2$ where $d = |x_1|$ is the distance by which the spring is compressed.

a) For $k = 10$ N/m, as in the example, calculate the work W done on the ball for spring compression distances of 0, 2, 4, 6, 8, and 10 cm. Plot your values on the axes below, provide an appropriate numerical scale on the vertical axis, then draw a smooth curve through the plotted points.

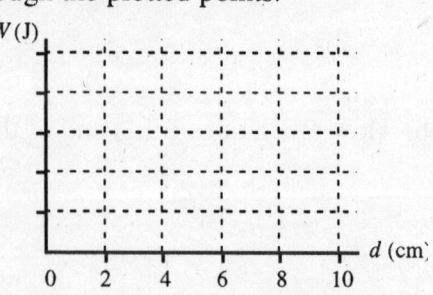

b) What geometric shape does this graph have? _____

c) Use your graph to determine how much work is done by the spring in pushing the ball through the 2 cm interval from $x = 2$ cm to $x = 0$ cm?

d) Use your graph to determine how much work is done by the spring in pushing the ball through the 2 cm interval from $x = 10$ cm to $x = 8$ cm?

e) Give an explanation as to why your answers to c) and d) are different.

9.10 Power

29. a) If you lift a 1 kg book 1 m, how much work do you do on it?

b) How much power must you provide to lift the book in 1 s? In 10 s? In 0.1 s?

30. a) How long does it take a 60 W light bulb to use 1 J of energy?

b) With what speed would you need to lift a 2 kg mass (≈ 5 lb) to provide it with 60 W of lifting power?

31. In Example 9-14, you saw that a sprinter's power output at time t is given by $P = ma^2t$. If a sprinter wants to increase her acceleration by 5% *and* to sustain that acceleration for the same length of time that she now does, by what percentage will she have to increase her maximum power output? (This question refers to *any* sprinter. Don't use the specific numerical values of Example 9-14.)

Chapter 10

Potential Energy and Conservation

10.1   Interacting Particles

10.2   Where Did the Kinetic Energy Go? Stored Energy

1. Describe each of the following situations first from a force-acceleration perspective and then again from a work-energy perspective.

a) A falling rock.

b) Lifting a book from the floor and placing it on a shelf.

c) Pushing a plastic ball into a spring-loaded gun.

2. Give a *specific* example of a situation in which:

a) $W_{ext} \to K$ with $\Delta U = 0$.

b) $W_{ext} \to U$ with $\Delta K = 0$.

c) $K \to U$ with $W_{ext} = 0$.

d) $U \to K$ with $W_{ext} = 0$.

e) $K \to W_{ext}$ with $\Delta U = 0$.

f) $U \to W_{ext}$ with $\Delta K = 0$.

10.3 Gravitational Potential Energy

3. A ball rolls up an inclined plane, then back down. What is the sign (+ or −) of:

	Rolling up	Rolling Down
Δy	_____	_____
$(F_{grav})_y$	_____	_____
W_{grav}	_____	_____
ΔU_{grav}	_____	_____

4. A particle moves in a vertical plane along a *closed* path, starting at A and eventually returning to its starting point. How much work is done on the particle by gravity? Explain.

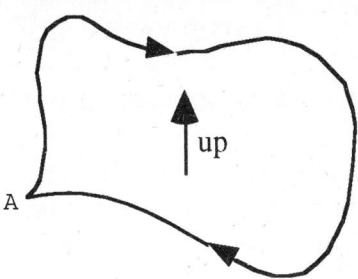

5. A roller coaster car rolls down a frictionless track, reaching speed v_f at the bottom.

a) If you want to car to go twice as fast at the bottom, by what factor must you increase the height of the hill?

b) Does your answer to a) depend on whether the track is straight or not? Explain.

6. The figure below shows a 1 kg object that is initially 1 m above the ground and rises to a height of 2 m above the ground. Allan, Bill, and Charles each measure its position, but each of them uses a different coordinate system. Fill in the table to show the initial and final potential energies as well as ΔU, as measured by our three aspiring scientists.

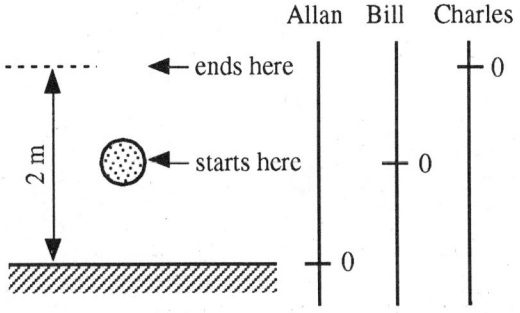

	U_i	U_f	ΔU
Allan			
Bill			
Charles			

7. Does a negative potential energy make sense? How can an object have a negative potential for converting stored energy into kinetic energy? Explain.

106   Chapter 10 Potential Energy and Conservation

10.4 Conservative and Nonconservative Forces

10.5 Conservation of Mechanical Energy

8. In the text we have found both the work-kinetic energy theorem $W = \Delta K$ and also the definition of potential energy $\Delta U = -W$. What is the relationship between these two statements? Are they saying the same thing, or something different?

9. Three balls of equal mass are fired simultaneously from the same height above the ground and with *equal* speeds. Ball 1 is fired straight up, ball 2 is fired straight down, and ball 3 is fired horizontally. Compare their speeds as they hit the ground. Explain your reasoning.

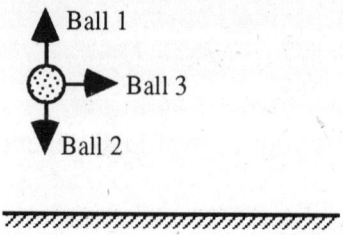

10. Below are shown three frictionless tracks. A ball is released from rest at the position shown on the left. To which point does it make it on the right before reversing directions and rolling back? Point B is the same height as its starting position.

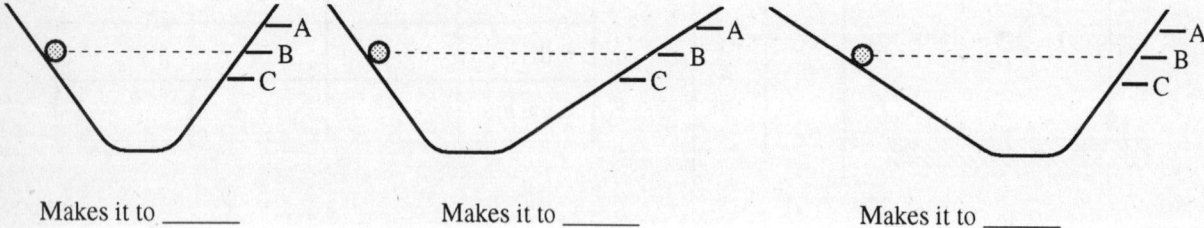

Makes it to _____   Makes it to _____   Makes it to _____

10.6 Elastic Potential Energy

11. A heavy object is released from rest at position 1 above a spring. It falls, contacts the spring at position 2, and finally comes to rest at position 3. Fill in the table below to indicate whether each of the quantities are +, –, or 0 during the intervals 1→2, 2→3, and 1→3.

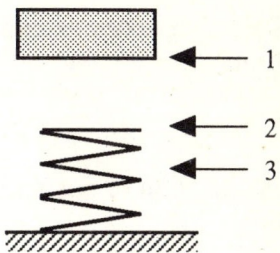

	1→2	2→3	1→3
ΔK			
ΔU_{grav}			
ΔU_{spring}			

12. A spring gun shoots out a plastic ball at speed v_0. The spring is then compressed twice the distance it was on the first shot.

a) By what factor is the spring's potential energy increased?

b) By what factor is the work needed to compress the spring increased?

c) By what factor is the ball's velocity increased?

10.7 Energy Diagrams

13. a) If the force on a particle is zero at some point in space, must its potential energy also be zero at that point? Explain.

b) If the potential energy of a particle is zero at some point in space, must the force on it also be zero at that point? Explain.

14. The figure shows the potential energy curve of a particle. Answer the following questions.

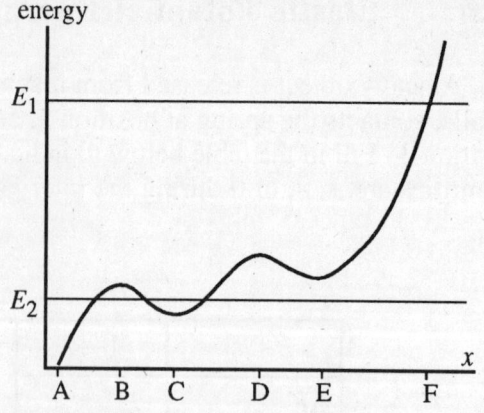

a) What position or positions are points of stable equilibrium?

b) What position or positions are points of unstable equilibrium?

c) Suppose the particle is at position x_A and moving to the right with total energy E_1. Describe its subsequent motion until it leaves the range of x shown on the axis. Your description should say where the particle is speeding up, slowing down, moving at steady speed, and turning around.

d) For a particle that has total energy E_2, what are the possible motions and where to they occur along the x-axis?

15. The graph below shows the potential energy curve of a particle moving along the x-axis under the influence of a conservative force.

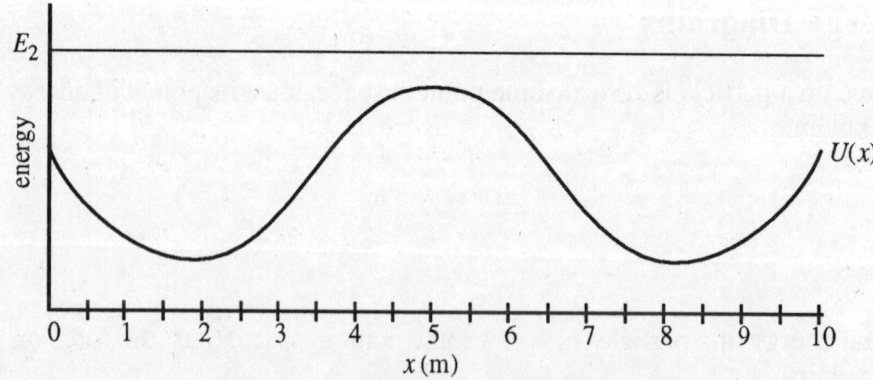

a) In which intervals of x is the force on the particle to the right?

Chapter 10 Potential Energy and Conservation 109

b) In which intervals of x is the force on the particle to the left?

c) At what value or values of x is the magnitude of the force a maximum?

d) What value or values of x are positions of stable equilibrium?

e) What value or values of x are positions of unstable equilibrium?

f) If the particle is released from rest at $x = 0$, will it reach $x = 10$ m? Explain.

g) Draw and label *on the graph* a total energy line E_1 for a particle that undergoes oscillations about one of the stable equilibrium points you identified in d).

16. Suppose the particle in the previous question were moving to the right at $x = 0$ with total energy E_2. The E_2 total energy line is shown on the graph.

a) At what value or values of x is the particle's speed a maximum?

b) At what value or values of x is the particle's speed a minimum?

c) At what value or values of x is the potential energy a maximum?

d) Does this particle have a turning point in the range of x covered by the graph? If so, where?

17. Below are a set of axes on which you are going to draw a potential energy curve. After appropriate experiments, you find the following information:

   A particle of energy E_1 oscillates between positions x_D and x_E.
   A particle of energy E_2 oscillates between positions x_C and x_F.
   A particle of energy E_3 oscillates between positions x_B and x_G.
   A particle of energy E_4 enters from the right, bounces at x_A, then never returns.

Draw a potential energy curve that is consistent with this information.

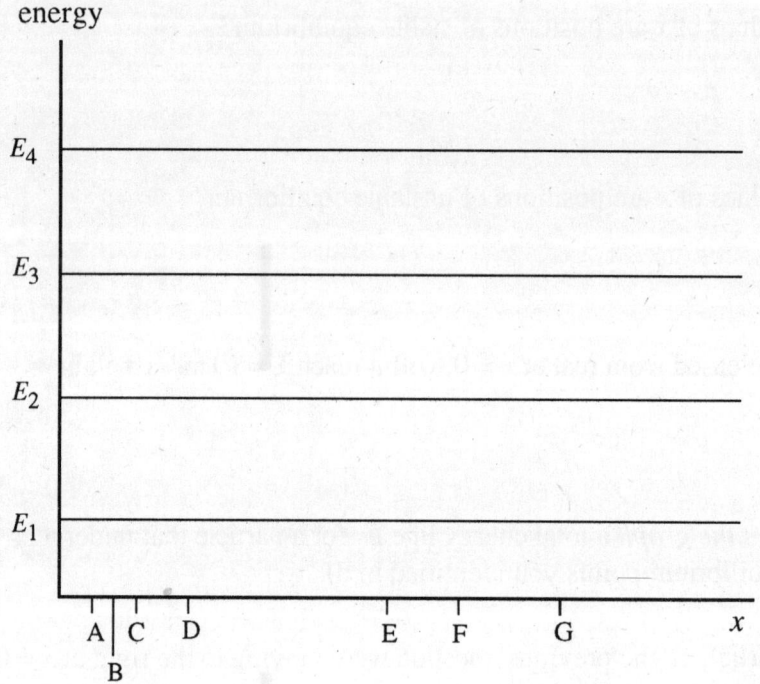

Chapter 11

Expanding the Concept of Energy

11.1 Forces That Do No Work

1. A spring of mass m is pushed down against the top of table, then it is released and it shoots up.

a) What force or forces are exerted *on* the spring?

b) Which of these forces do work and which do no work? Explain.

c) For any forces that you listed in b) as doing no work, give an example of a situation in which they *would* do actual work on the spring.

11.2 Microphysics and Macrophysics

2. A closed box with a few marbles inside is thrown across the room.

a) Does the box as a whole have kinetic energy? _____

b) Do the marbles inside have kinetic energy? _____

c) What are the similarities and the differences between the *macro* kinetic energy of the box as a whole and the *micro* kinetic energy of the marbles inside?

11.3 The Center of Mass

11.4 The Energy Equation

3. A firefighter slides down a fire pole at constant speed.
a) Draw a free body diagram of the firefighter and identify all the forces acting on him.

b) Which of these forces do work on the firefighter? _____

c) Which of these forces can be associated with a potential energy? _____

d) There are three possible choices of "the system:" i) firefighter, ii) firefighter+pole, or iii) firefighter+pole+earth. Which of these is the best choice for using energy conservation? Why?

e) Write down the energy equation for the system you chose in d).

f) *Interpret* your equation. That is:
   i) What is the initial form of the energy?

   ii) What is the final form of the energy?

   iii) What has happened as a consequence of the energy transfer?

g) Has energy been converted into heat? Why or why not?

11.5 The Law of Conservation of Energy

4. a) What is the distinction between E_{therm} and E_{int}?

   b) What is the distinction between *heat* and *work*?

   c) What is the distinction between *heat* and *thermal energy*?

5. A car uses a certain amount of gasoline to accelerate from rest to 5 m/s. Compared to the gasoline used to go from 0 to 5 m/s, how much gasoline is required to go from 5 to 10 m/s? (The speeds are slow enough to neglect energy losses to air resistance.) Give a physics explanation for your answer.

114   Chapter 11 Expanding the Concept of Energy

For Exercises 6 - 9, describe the forces and energy transfers involved in the motion. In particular:
   i) Identify the force or forces that accelerate or decelerate the object.
   ii) Identify any forces that do zero-work.
   iii) Identify the energy transfers involved in the motion.
Your descriptions should be similar to those at the end of Examples 11-2 and 11-3 (but without the numbers). Pictures and diagrams will be helpful.

6.   A person wearing rollerblades pushes off backwards from a wall.

7.   A horizontally-moving ball of putty collides with a wall and sticks to it.

8.   A woman runs up a flight of stairs.

9. A piston is pushed inward to compress a cylinder of air.

11.6 Power Revisited

10. A boy on frictionless rollerblades bends his arms and pushes off backwards from a wall. He puts on a lead vest that doubles his weight, then pushes off again with the same pushing force and arm bending. Compared to his first push:

a) By how much has the work done on him changed?

b) By how much has his final speed (as his fingers leave the wall) changed?

c) By how much has ΔE_{chem} changed?

d) By how much has the time Δt it takes to push off changed?

e) By how much has his maximum power output changed?

11.7 Collisions Revisited

No exercises.

Chapter 12

Newton's Theory of Gravity

12.1   A Little History

12.2   Isaac Newton

12.3   Newton's Law of Gravity

1. How does the magnitude of the earth's gravitational force on the sun compare to the magnitude of the sun's gravitational force on the earth? Larger, smaller, or the same? Why?

2. In a binary star system (two stars circling each other) Star A is three times as massive as Star B.
a) Compare the magnitude of the gravitational force on Star A to the gravitational force on Star B.

b) Compare the acceleration of Star A to the acceleration of Star B.

3. Comets orbit the sun in highly elliptical orbits. A new comet is sighted on Day 1.

a) On Day 30, the comet's acceleration a_{30} is observed to be twice as large as its acceleration a_1 was on Day 1. How does the comet's distance from the sun r_{30} on Day 30 compare to its distance r_1 on Day 1?

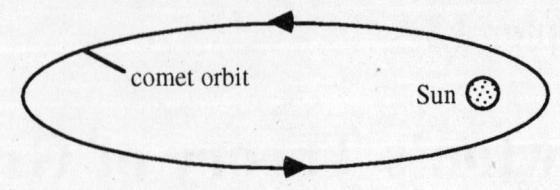

b) On Day 60 the comet has rounded the sun and is headed back out to the farthest reaches of the solar system. The force F_{60} on the comet is the same on this day as the force F_{30} was on Day 30, but the comet's distance from the sun r_{60} is only 90% of its distance on Day 30. Astronomers recognize that the comet has lost mass -- it was "boiled away" by the heat of the sun during the time of closest approach, between Days 40 and 50, and formed the comet's tail. What percentage of its initial mass did the comet loose?

12.4 Little g and Big G

4. Explain why the Space Shuttle astronauts are "weightless."

5. How far away from the earth does an orbiting spacecraft have to be in order for the astronauts inside to be weightless?

6. Don't do any calculations, but *describe* a method by which you could "weigh the sun."

12.5 Gravitational Potential Energy

7. Give a physics explanation of *why* the potential energy of two masses is negative. (Note: Saying "because that's what the formula gives" is *not* an explanation. An *explanation* requires going back to the basic ideas and definitions of force and potential energy.)

12.6 Satellite Orbits and Energies

8. Planet X, circling the star Omega, has a "year" that is 200 earth days long. Planet Y circles Omega at twice the distance of Planet X. How long is a year on Planet Y?

9. a) When the Space Shuttle wants to return to earth from a circular orbit, in which direction does it fire its rocket engine? Why?

b) Make a drawing, similar to Fig. 12-16, showing the earth, the Shuttle's orbit before firing its rocket engine, and its new orbit after firing its rocket engine.

c) Suppose the earth had no atmosphere, so the Shuttle would continue unimpeded along its new orbit until intersecting the ground (ouch!). As it descends, would its speed increase, decrease, or stay the same? Explain your answer in terms of energy transfer.

Chapter 13

Oscillations

13.1 Periodic Motion

1. Give three examples of *oscillatory* motion. (Note that *circular* motion is not the same thing as oscillatory motion.)

2. On the axes below, sketch several cycles of the displacement-versus-time graph for:
a) A particle undergoing symmetric periodic motion but that is *not* SHM.

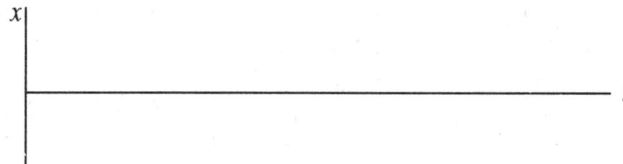

b) A particle undergoing asymmetric periodic motion.

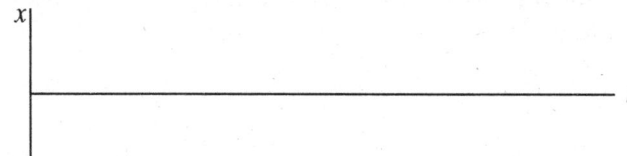

c) A particle undergoing simple harmonic motion.

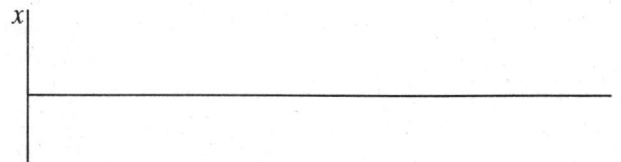

d) A particle whose motion is oscillatory but *not* periodic.

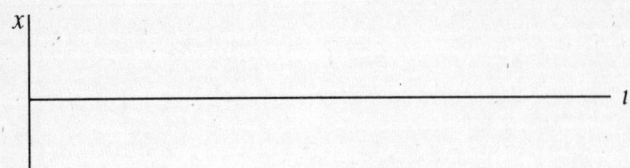

3. Consider the particle whose motion is represented by the *x*-versus-*t* graph below.

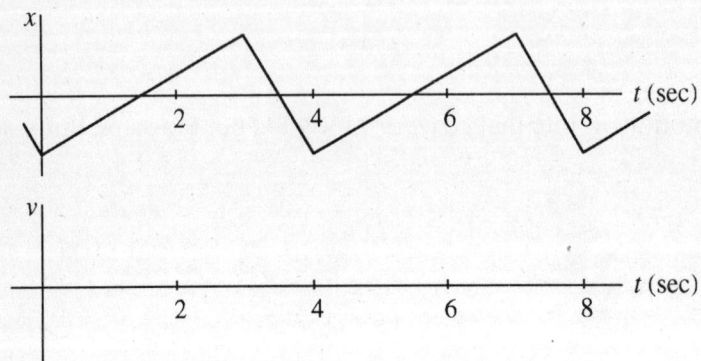

a) Is this periodic motion? _____ b) Is this motion SHM? _____

c) What is the period of the motion? _____
d) What is the frequency of the motion?

e) Recalling what you learned in Chapter 3 about relating velocity graphs to position graphs, draw the corresponding velocity-versus-time graph for $0 \leq t \leq 8$ s. Make sure the vertical heights of your graph properly indicate where the motion is faster and slower.

4. Shown below is the velocity-versus-time graph of a particle.

a) What is the period of the motion? _____

b) Draw the corresponding position-versus-time graph for $0 \leq t \leq 12$ s. Show the particle traveling equal distances in the positive and negative directions.

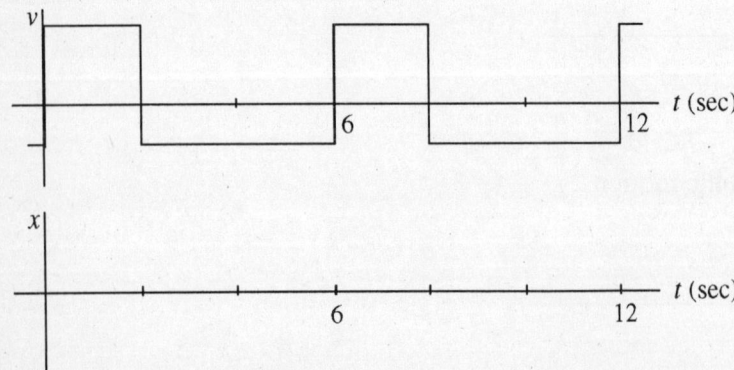

5. Graph a) shows a position-versus-time graph that is parabolic: $y = ct^2$ where c is a constant. Graph b) shows a position-versus-time graph that is an inverted parabola: $y = -ct^2$. For each:
   i) Draw a graph of velocity-versus-time, then
   ii) Write an exact expression for the velocity v as a function of time.
Make sure your graphical and algebraic answers agree!

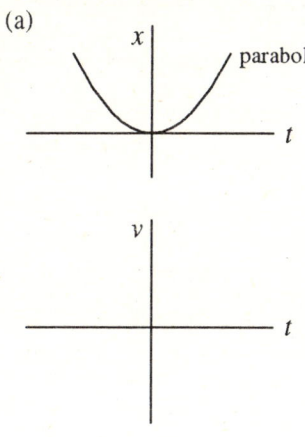

v = _____

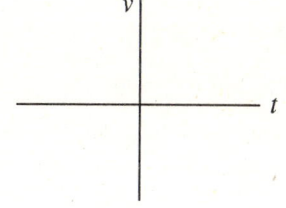

v = _____

6. The graph below is the position-versus-time graph of an oscillating particle. It is constructed of *parabolic* segments, alternating up and down and joined at $x = 0$.

a) Is this simple harmonic motion? Why or why not?

b) Draw the corresponding velocity-versus-time graph.
c) Draw the corresponding acceleration-versus-time graph.

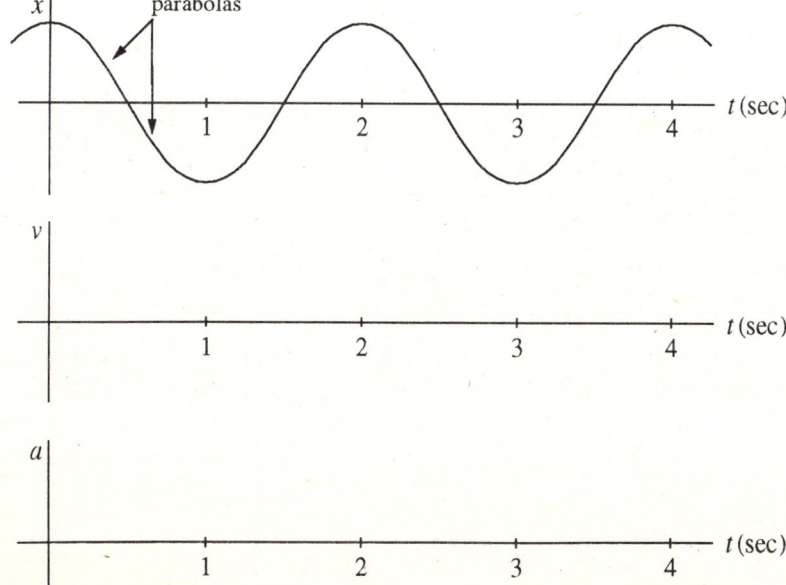

d) At what times is the displacement a maximum? _____

At those times, is the velocity a maximum, a minimum, zero, or in-between? _____

At those times, is the acceleration a maximum, a minimum, or zero? _____

e) At what times is the displacement a minimum (most negative)? _____

At those times, is the velocity a maximum, a minimum, zero, or in-between? _____

At those times, is the acceleration a maximum, a minimum, or zero? _____

f) At what times is the velocity a maximum? _____

At those times, where is the particle? _____

g) At what times is the velocity a minimum (most negative)? _____

At those times, where is the particle? _____

h) Your answers to "where is the particle?" should have been the same for f) and g). How can v be both a maximum *and* a minimum when the particle is at the same place? Explain.

i) Is there a simple relationship between the *sign* of the displacement and the *sign* of the acceleration? If so, what is it?

13.2 Simple Harmonic Motion

7. A particle goes around a circle 5 times at constant speed, taking a total of 2.5 seconds.

a) Through what angle *in degrees* has the particle moved? _____

b) Through what angle *in radians* has the particle moved? _____

c) What is the particle's frequency f?

d) What is the particle's angular frequency ω? Determine ω directly from your answer to b).

e) Does $\omega = 2\pi f$? _____

f) Do f and ω have the same units or different units? Explain.

8. A particle moves counterclockwise around a circle at constant speed. For each of the phase constants given below:
   i) Show with a dot *on the circle* the particle's starting position.
   ii) Sketch 3 cycles of the particle's *x*-versus-*t* graph.

   a) $\phi_0 = 0$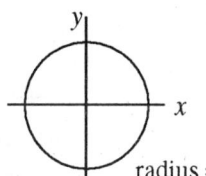

   b) $\phi_0 = \pi/2$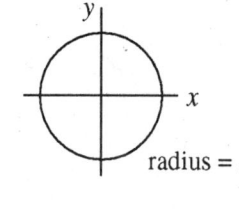

   c) $\phi_0 = \pi$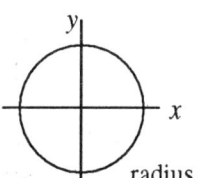

   d) $\phi_0 = -\pi/2$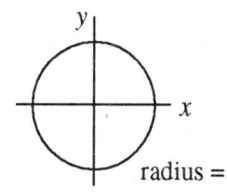

9. a) On the first set of axes below sketch three cycles of the *x*-versus-*t* graphs for a particle moving around a circle at constant speed with phase constants of i) $\phi_0 = \pi/2$ and ii) $\phi_0 = -\pi/2$.

   b) By observing the slopes of your first graphs, use the second set of axes to sketch velocity-versus-time graphs (*v*-versus-*t*) for the particles. Make sure each velocity graph aligns vertically with the correct points on the *x*-versus-*t* graph.

   i) $\phi_0 = \pi/2$

   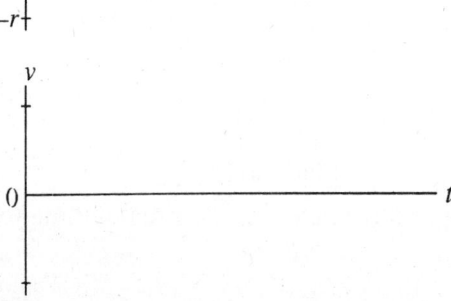

   ii) $\phi_0 = -\pi/2$

   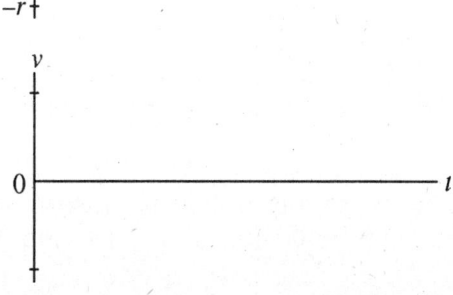

126 Chapter 13 Oscillations

10. The graph below represents a particle moving counterclockwise around a circle at constant speed.

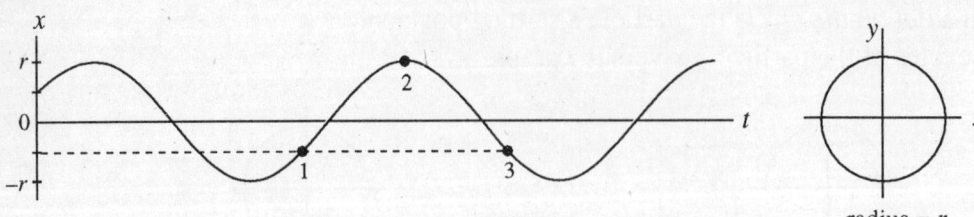

a) What is the phase constant ϕ_0? Explain how you determined it.

b) Note on the graph that x gets more positive immediately after $t = 0$. Is this true for a particle moving counterclockwise on a circle with the value of ϕ_0 you answered in a)

c) What is the phase of the particle at each of the three numbered points on the graph?

Phase at 1:_____ Phase at 2:_____ Phase at 3:_____

d) On the circle next to the graph above, use dots to show the position of the particle at the times corresponding to points 1, 2, and 3 on the graph. Label each dot with the appropriate number.

11. The graph shows *velocity*-versus-time for a particle moving counterclockwise around a circle.

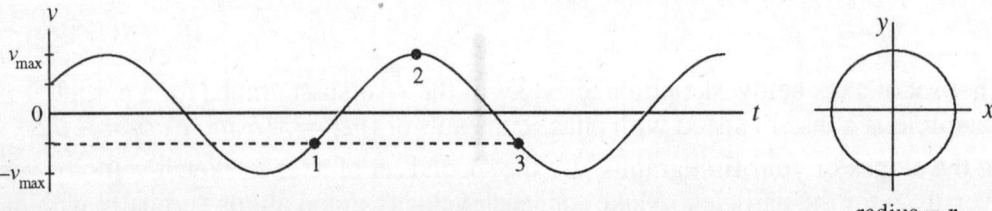

a) What is the phase constant ϕ_0? Explain how you determined it.

b) What is the phase of the particle at each of the three labeled points on the graph?

Phase at 1:_____ Phase at 2:_____ Phase at 3:_____

c) On the circle above, use dots to show the *position* of the particle at the times corresponding to points 1, 2, and 3 on the graph. Label each dot with the appropriate number.

13.3 The Equation of Motion

12. A block oscillating on a spring has a frequency $f = 2$ Hz.

a) What is the frequency if the block's mass is doubled? (Note: You do not know values for either m or k. Do *not* assume any particular values for them. The required analysis involved thinking about *ratios*.) Explain your reasoning; show your work.

b) What is the frequency if the value of the spring constant is quadrupled?

c) What is the frequency if the block's mass *and* the value of the spring constant are both doubled?

d) What is the frequency if the block's mass is doubled *and* the value of the spring constant is halved?

e) What is the frequency if the oscillation amplitude is doubled while m and k are unchanged?

13. The graph at the right is the position-versus-time graph for a simple harmonic oscillator.

a) Draw the v-versus-t and a-versus-t graphs for the oscillator on the axes provided. Make sure that each aligns vertically with the corresponding points on the x-versus-t graph.

b) When x is greater than zero, is a ever greater than zero? If so, at which points in the cycle?

c) When x is less than zero, is a ever less than zero? If so, at which points in the cycle?

d) Can you make a general conclusion about the relationship between the sign of x and the sign of a?

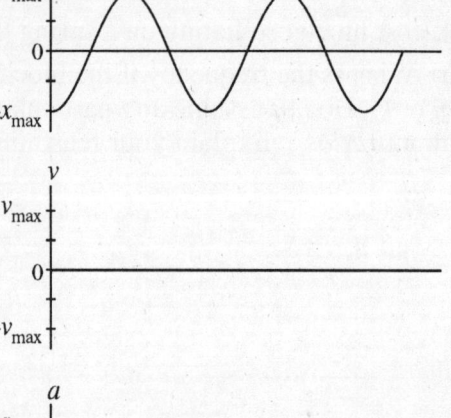

e) When x is greater than zero, is v ever greater than zero? If so, describe physically how the oscillator is moving at those times.

14. For the oscillation shown on the left below:
a) What is the phase constant ϕ_0?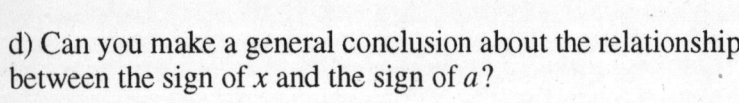
b) Use dots to mark *on the graph* the points at which the phase of the oscillator is $\pi/2$.
c) Sketch the corresponding v-versus-t graph on the axes below the x-versus-t graph.
d) On the axes at the right, sketch two cycles of the x-versus-t and the v-versus-t graphs if the value of ϕ_0 found in a) is replaced by its negative, $-\phi_0$.

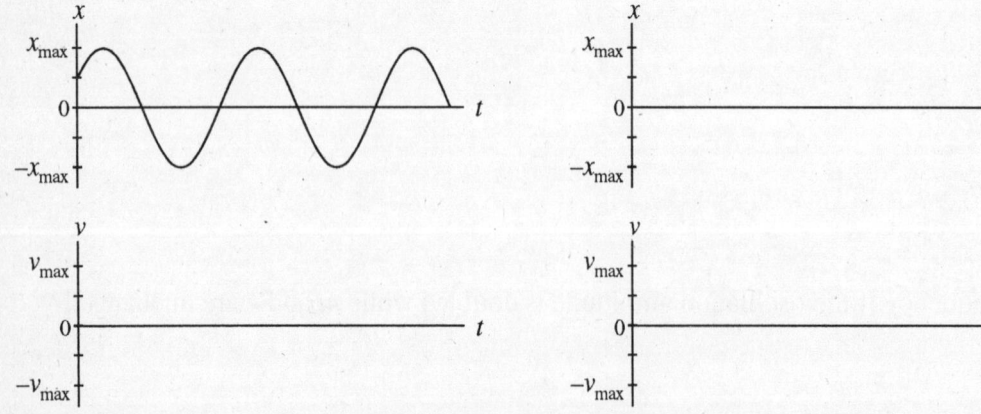

d) Describe *physically* what is the same and what is different about the initial conditions for two oscillators having "equal but opposite" phase constants ϕ_0 and $-\phi_0$.

15. The top graph is the position-versus-time for a block oscillating on a spring. On the axes below, sketch the position-versus-time graph for this block for the following situations:

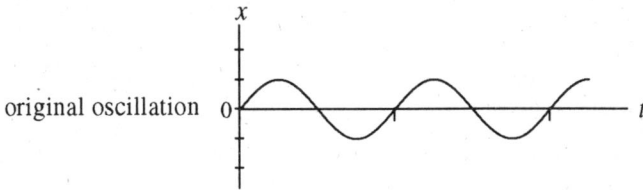

a) The amplitude and the frequency are doubled.

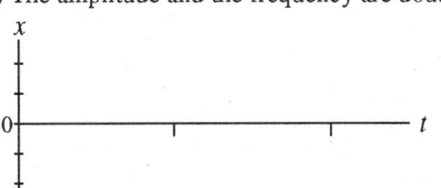

c) The phase constant is increased by $\pi/2$.

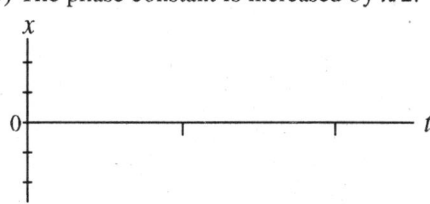

b) The amplitude is halved and the mass is quadrupoled.

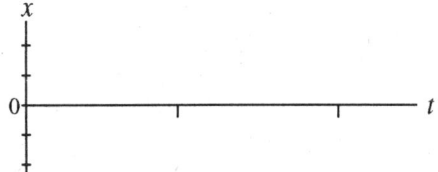

d) The maximum speed is doubled while the amplitude remains constant.

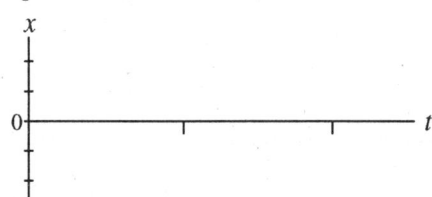

NOTE: The changes described in each part refer back to the original oscillator, not to the oscillator of the previous part of the question. Assume that all other parameters remain constant. Use the same horizontal and vertical scales as the original oscillation graph.

16. For a) and b), determine
    i) The angular frequency ω.
    ii) The oscillation amplitude A.
    iii) The phase constant ϕ_0.
Note that a) and b) are independent questions. b) is *not* the velocity graph of a).

a)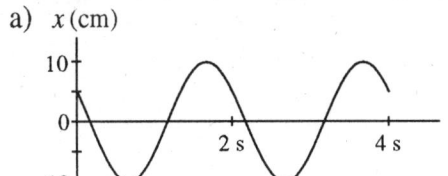

b)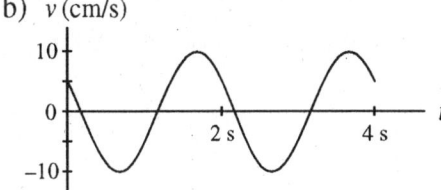

13.4 Vertical Oscillations

13.5 The Pendulum

17. A pendulum on Planet X, where the value of g is unknown, oscillates with a period of 2 second. What is the period of this pendulum if
NOTE: You do not know the values of m, L, or g so do not assume any specific values.
a) Its mass is doubled?

b) Its length is doubled?

c) Its oscillation amplitude is doubled?

18. The graph shows displacement s-versus-time for an oscillating pendulum.

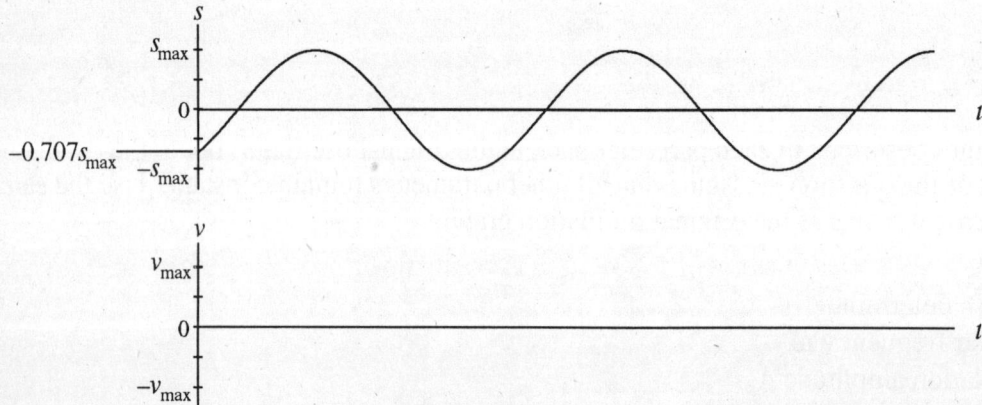

a) Draw the pendulum's velocity-versus-time graph.
b) What is the value of the phase constant ϕ_0?

c) In the space at the right, draw a *picture* of the pendulum that shows (and labels!)
   i) The extremes of its motion,
   ii) Its position at $t = 0$, and
   iii) Its direction of motion (using an arrow) at $t = 0$.

13.6 Energy in Oscillating Systems

19. The figure shows the potential energy diagram for an oscillator. Also shown is the total energy line of a particle.

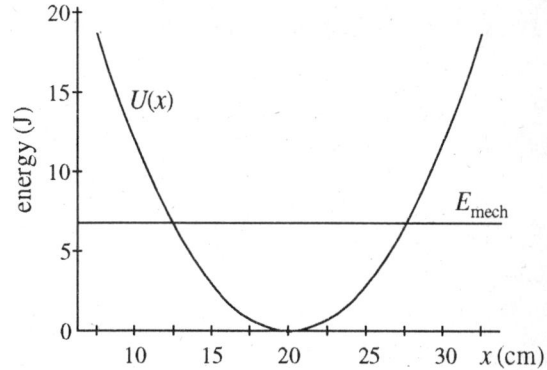

a) Where are the turning points of the motion? Explain how you identify them.

b) What is the particle's maximum kinetic energy?

c) Draw a graph of the particle's kinetic energy as a function of position.

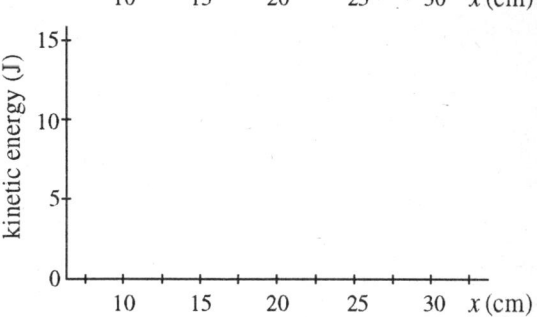

d) What will be the turning points if the particle's total energy is doubled?

20. A block oscillating on a spring has an amplitude of 20 cm. What will the block's amplitude be if its energy is doubled? Explain.

21. A block oscillating on a spring has a maximum speed of 20 cm/s. What will the block's maximum speed be if its energy is doubled? Explain.

22. The figure shows the potential energy diagram of a particle.

a) Is the particle's motion periodic? How can you tell?

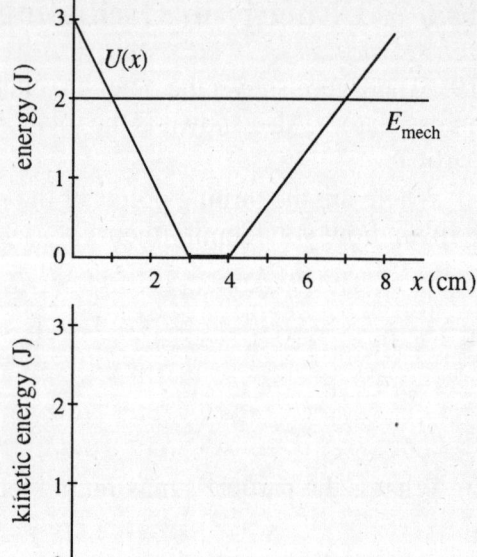

b) Is the particle's motion simple harmonic motion? How can you tell?

c) What is the amplitude of the motion?

d) Sketch a graph of the particle's kinetic energy as a function of position.

23. Equation 16-34 in the text states that $\frac{1}{2}kx_{max}^2 = \frac{1}{2}mv_{max}^2$. What does this mean? Write a couple of sentences explaining how to interpret this equation.

13.7   Damped Oscillations

24. If the damping constant b of an oscillator is increased,

a) Is the medium more resistive or less resistive? _____

b) Do the oscillations damp out more quickly or less quickly? _____

c) Is the time constant τ increased or decreased? _____

25. A block on a spring oscillates *horizontally* on a table *with* friction. Draw force vectors on the block, showing *all* forces. Use vectors of appropriate lengths.

a) The mass is to the left of the equilibrium point and approaching it.

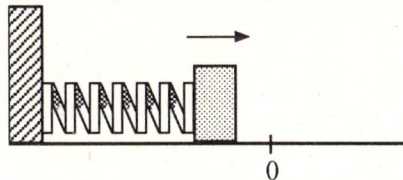

b) The mass is to the right of the equilibrium point and moving away from it.

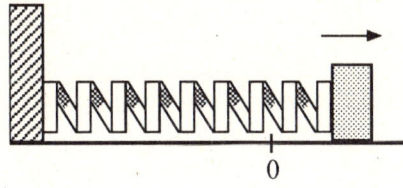

c) The mass is to the right of the equilibrium point and approaching it.

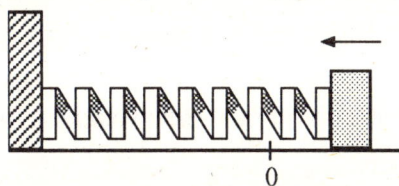

d) Draw and label the block's acceleration vector $\vec{a}$ under each of your three diagrams.

26. An oscillator has a period $T = 1$ s. It is released from rest at $x = x_{max}$ at $t = 0$. Sketch the position-versus-time graph for $0 \leq t \leq 5$ s for the case that:

a) $\tau = 2T$

b) $\tau = T$

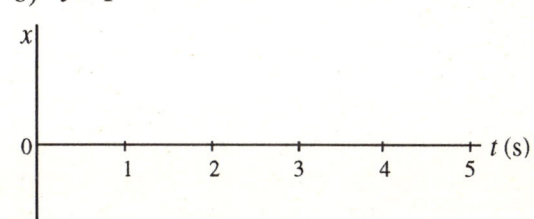

c) $\tau = (1/2)T$

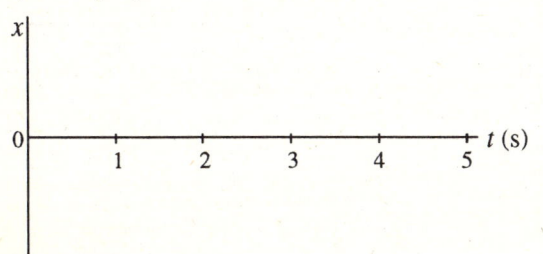

d) Oscillation envelopes

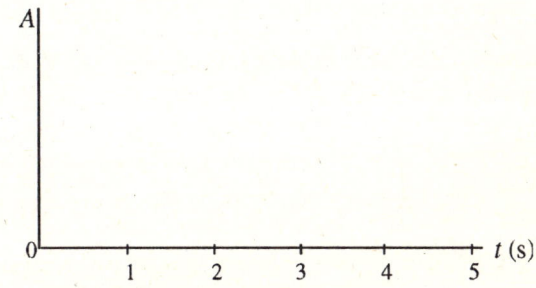

d) Now draw and label each of the three oscillation envelopes on the fourth set of axes.

27. a) Describe the difference between τ and T. Don't just *name* them -- the name is arbitrary -- but say what is different about the physical concepts that they represent.

b) Describe the difference between τ and $t_{1/2}$.

13.8 Driven Oscillations and Resonance

28. What is the difference between the driving frequency and the natural frequency of an oscillator?

29. A car drives along a bumpy road on which the bumps are equally spaced. At a speed of 20 mph, the frequency of bumping is equal to the natural frequency of the car bouncing on its springs.

a) Draw a graph of the car's vertical bouncing amplitude as a function of its speed if the car has new shock absorbers (large damping coefficient).

b) Draw a graph of the car's vertical bouncing amplitude as a function of its speed if the car has worn out shock absorbers (small damping coefficient).

Place both graphs on the same axes, then label them as to which is which.

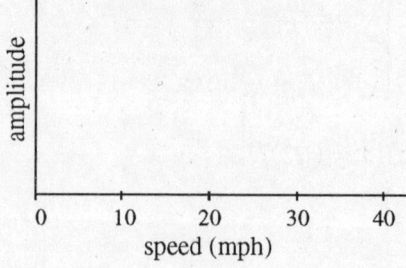

Chapter 14

Traveling Waves

14.1 Wave Physics

1. a) In your own words, define what a "transverse wave" is.

   b) Give an example of a wave that, from your own experience, you *know* is a transverse wave. Cite the evidence that tells you this is a transverse wave.

2. a) In your own words, define what a "longitudinal wave" is.

   b) Give an example of a wave that, from your own experience, you *know* is a longitudinal wave. Cite the evidence that tells you this is a longitudinal wave.

14.2 Traveling Waves

3. The figure shows a wave pulse on a string. The pulse is traveling to the right at 1000 cm/s. This figure shows the wave pulse at time $t = 0$. There is a small bead attached to the string at the point $x = 10$ cm. As the wave pulse passes by, the bead will be displaced upward and then will return to its equilibrium position.

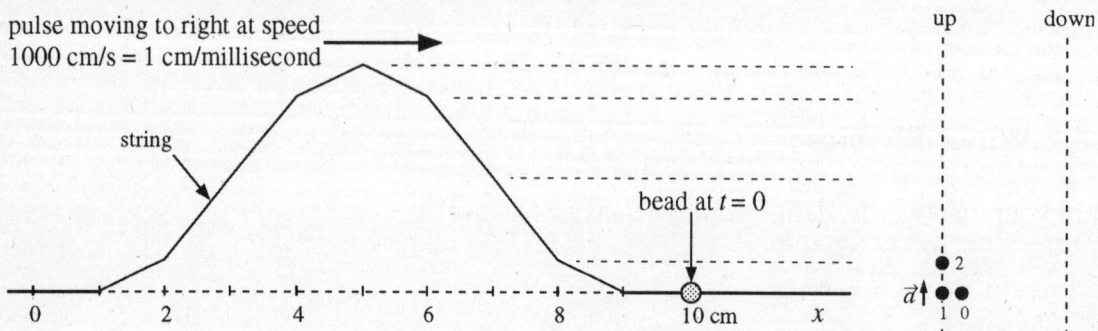

a) Use the two lines at the right to draw a motion diagram of the bead. Show one frame every millisecond, from $t = 0$ ms to $t = 10$ ms. Draw the frames showing upward motion of the bead on the left line and frames showing the downward motion of the bead on the right line. Show the very top position, at $t = 5$ ms, on *both* lines. Label each frame with the appropriate time, in milliseconds. The first three frames, at 0, 1, and 2 ms, are already shown and labeled.

b) Draw and label the $\vec{v}$ and $\vec{a}$ vectors on your motion diagram. Write $\vec{a} = 0$ if appropriate. Recall that velocity vectors go *between* the dots and acceleration vectors, which relate two velocity vectors, go *at* the dots. The acceleration vector at the $t = 1$ ms dot is already shown.

c) At each instant of time, the bead feels two tension forces $\vec{T}_R$ and $\vec{T}_L$ pulling to the right and to the left. Tension forces, recall, act along the direction of the string. Draw 11 free body diagrams below, from $t = 0$ ms to $t = 10$ ms, showing (with labels!) the two tension forces *and* the net force $\vec{F}_{net}$. Use a different color for $\vec{F}_{net}$. If $\vec{F}_{net} = 0$, write that beside the figure. The $t = 2$ ms drawing is done for you as an illustration. (Keep in mind that the bead moves in the vertical direction only. This tells you something about the direction of $\vec{F}_{net}$.)

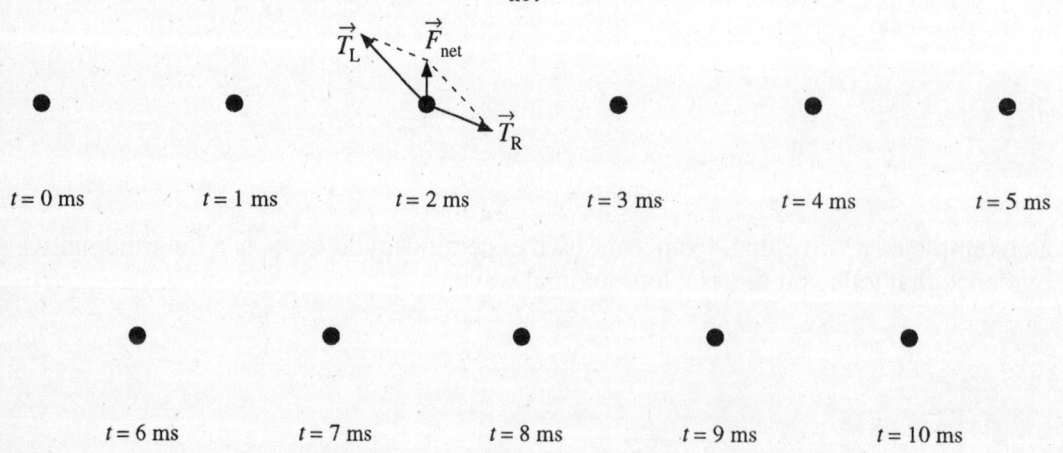

d) Compare your answers to b) and c). Is Newton's second law obeyed? That is, do $\vec{F}_{net}$ and $\vec{a}$ always point the same direction? If not, why not?

14.3 One-Dimensional Waves

4. A wave pulse travels along a string with a speed of 200 cm/s. What will the speed be if:
(Note: Each of the parts below is independent of the others and refers to changes made to the original string. All other parameters of the string are held constant.)

a) The string's tension is doubled?

b) The string's mass is quadrupled (but its length is unchanged)?

c) The string's length is quadrupled (but its mass is unchanged)?

d) The string's mass and length are both quadrupled?

5. The graph is a history graph — displacement as a function of time at *one* point on a string. Did the displacement at this point reach its maximum of 2 mm *before* or *after* the interval of time when it had a constant 1 mm displacement? Explain how you interpreted the graph to answer this question.

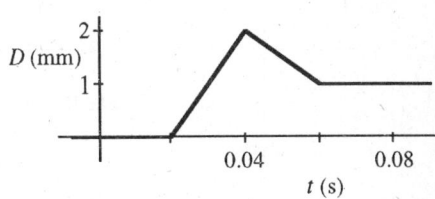

138　Chapter 14 Traveling Waves

6. a) The figure on the left below is a snapshot graph $D(x, t = 0)$, at time $t = 0$, of a wave pulse on a string. The pulse is traveling to the right at a speed of 100 cm/s. We will always assume in exercises like this that the string extends very far in both directions, even if not shown on the graph, and that the pulse has been traveling for a very long time. On the axes provided, draw snapshot graphs of the wave pulse at times

　　i)　$t = -0.01$ s　　　ii)　$t = 0.00$ s　　　iii)　$t = 0.01$ s
　　iv)　$t = 0.02$ s　　　v)　$t = 0.03$ s　　　vi)　$t = 0.04$ s

b) Repeat a) for the snapshot graph on the right of a pulse traveling to the left at 100 cm/s.

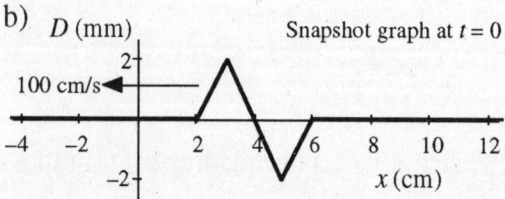
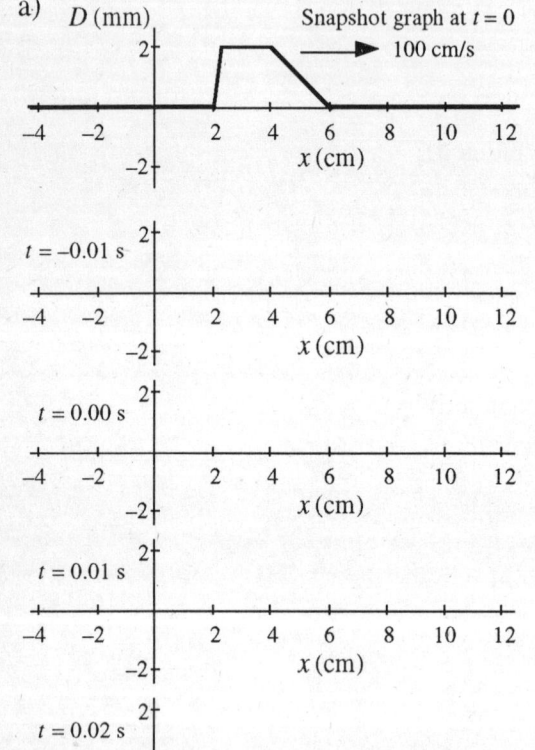
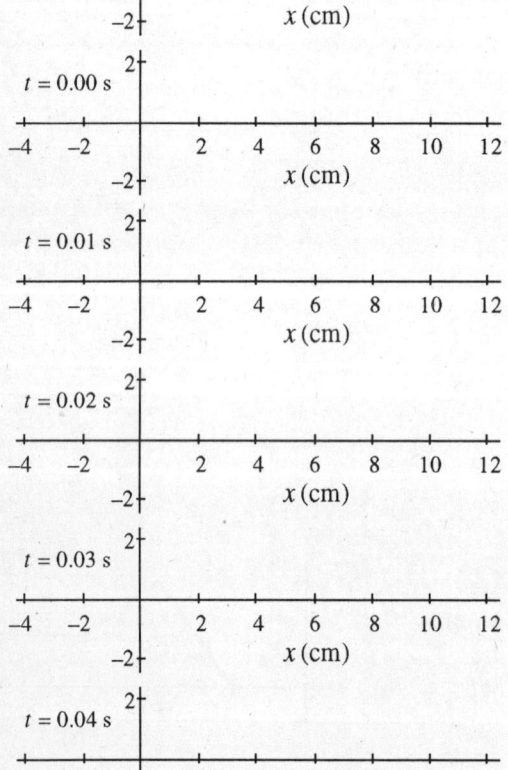
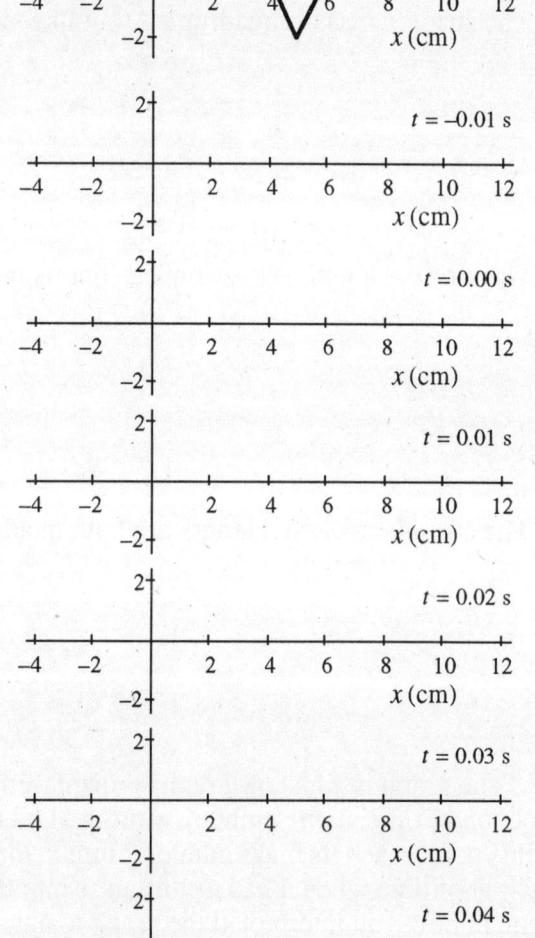
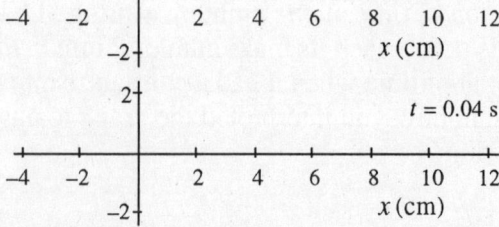

7. The snapshot graph shown here is reproduced from Exercise 5a.

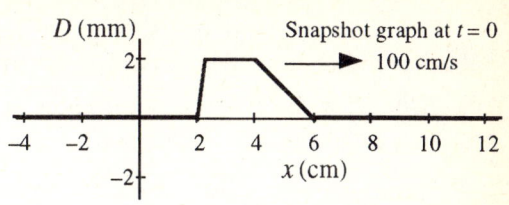

a) On the *middle* set of axes below, draw the *history* graph $D(x = 6$ cm, $t)$ showing the displacement at the point $x = 6$ cm as a function of time. You may want to refer to the graphs you drew for Exercise 5a to see what is happening at $x = 6$ cm at different instants of time.

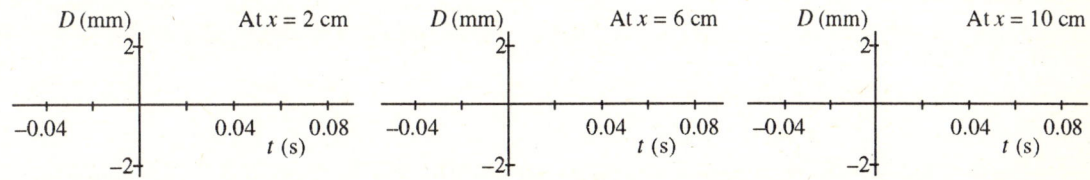

b) Now draw the history graphs $D(x = 2$ cm, $t)$ and $D(x = 10$ cm, $t)$ for the points $x = 2$ cm and $x = 10$ cm on the left and right sets of axes, respectively.

8. The snapshot graph shown here is reproduced from Exercise 5b.

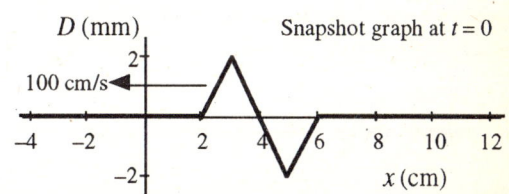

a) Draw the history graph $D(x = 0, t)$ for this wave at the point $x = 0$.

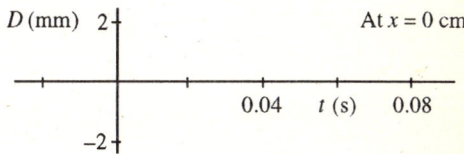

b) Referring to your displacement-versus-time history graph of a), draw the *velocity*-versus-time graph for this piece of the string. (Imagine, if you wish, painting a dot on the string at $x = 0$. What is the velocity of this dot at different instants of time as the wave passes by?)

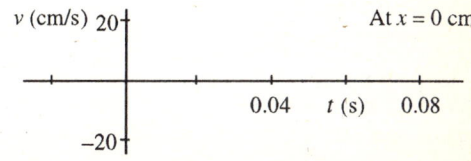

c) Give a description *in words* of how the piece of string at $x = 0$ moves during different intervals of time as the wave passes. That is, divide the motion into intervals — such as 0.01 s to 0.03 s — and say whether the string is moving up, moving down, or at rest. If it is moving, say what its speed is.

d) As a wave passes through a medium, is the speed of a particle in the medium the same as or different than the speed of the wave through the medium?

140   Chapter 14 Traveling Waves

9. Below are several snapshot graphs of wave pulses on a string. The speed and direction of each pulse are indicated. Note the time at which each snapshot was taken; they are not all shown at $t = 0$. For each, draw the history graph at the specified point on the x-axis. Note that no time scale is provided on the t-axis. Determine the appropriate time scale and label the t-axis appropriately.

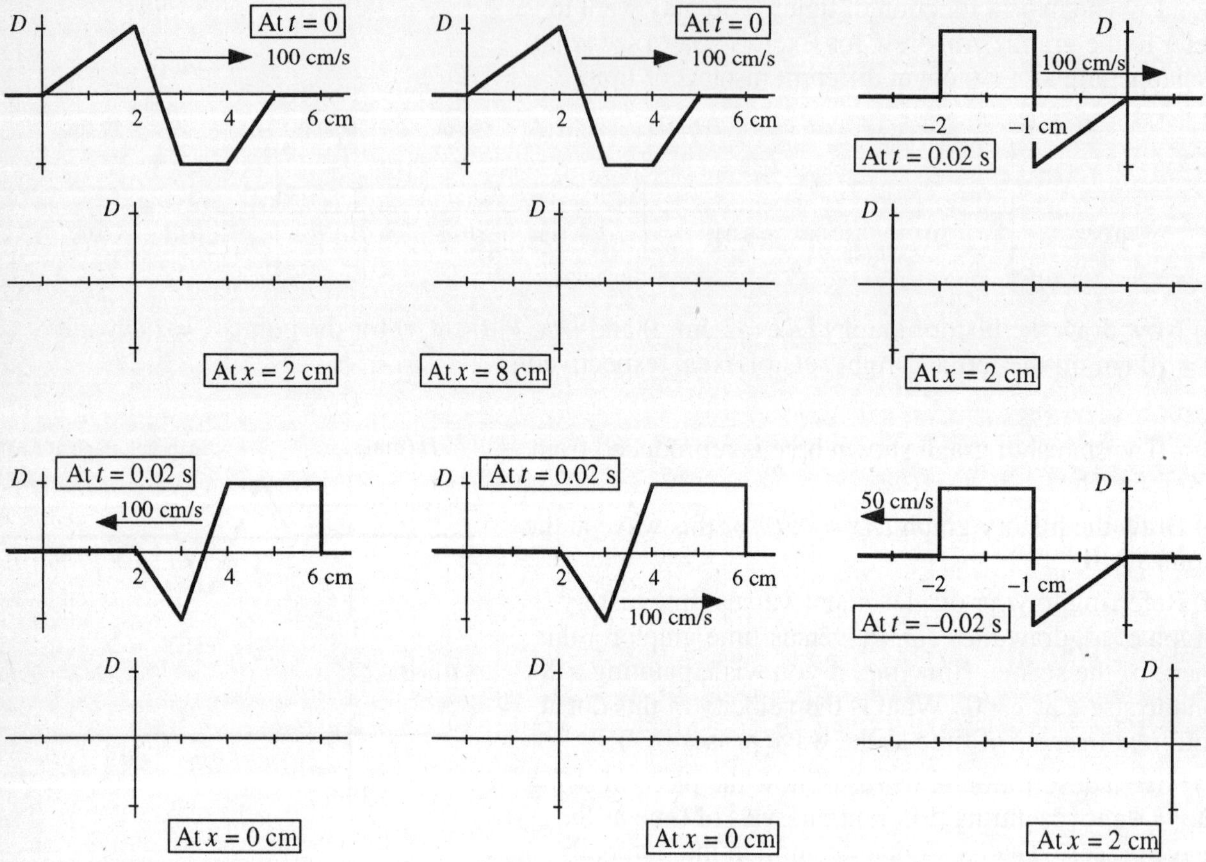

10. A history graph $D(x = 0, t)$ is shown for the $x = 0$ point on a string. The wave pulse is moving to the right with a speed of 100 cm/s.

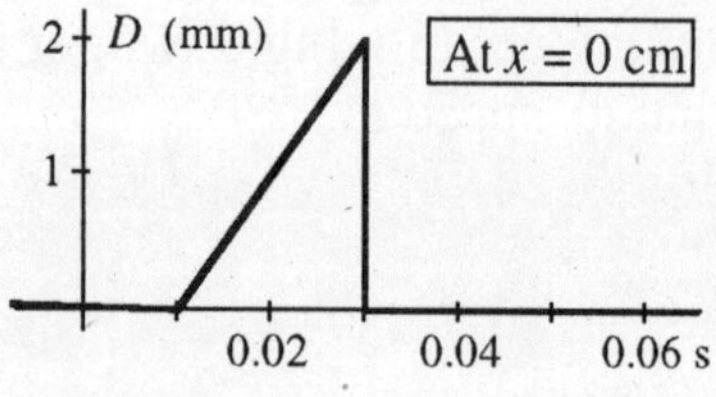

a) Does the $x = 0$ point on the string rise quickly then fall slowly, or rise slowly and then fall quickly? Explain how you tell this from the graph.

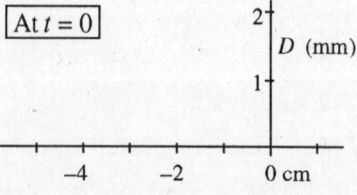

b) At what time does the wave pulse first arrive at $x = 0$?

c) At $t = 0$, how far is the leading edge of the wave pulse from $x = 0$? Explain your reasoning.

d) At $t = 0$, is the leading edge to the right or to the left of $x = 0$? _____

e) At what time does the wave pulse leave $x = 0$? _____

f) At $t = 0$, how far is the trailing edge of the pulse from $x = 0$? _____

g) At what time does the displacement at $x = 0$ first reach 1 mm? _____

h) At $t = 0$, how far from $x = 0$ is the point on the wave where $D = 1$ mm? _____

i) By referring to your answers to c), d), f), and h), draw (on the axes above) a snapshot graph $D(x, t = 0)$ showing the wave pulse on the string at $t = 0$.

11. Shown below are a history graph <u>and</u> a snapshot graph for a wave pulse on a string. They describe the same wave from two different perspectives.

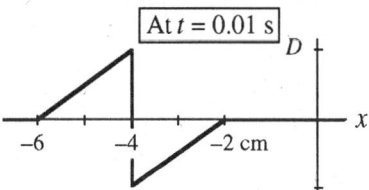
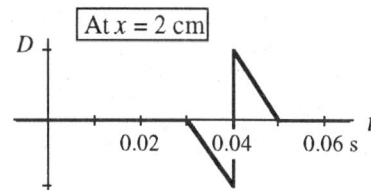

What is the speed of this wave, and in which direction is it traveling? Explain your reasoning.

12. Below are several history graphs for wave pulses on a string. The speed and direction of each pulse are indicated. Note the position at which each history was recorded; they are not all at $x = 0$.

For each, draw the snapshot graph at the specified instant of time. Note that no distance scale is provided on the x-axis. Determine the appropriate distance scale and label the x-axis appropriately.

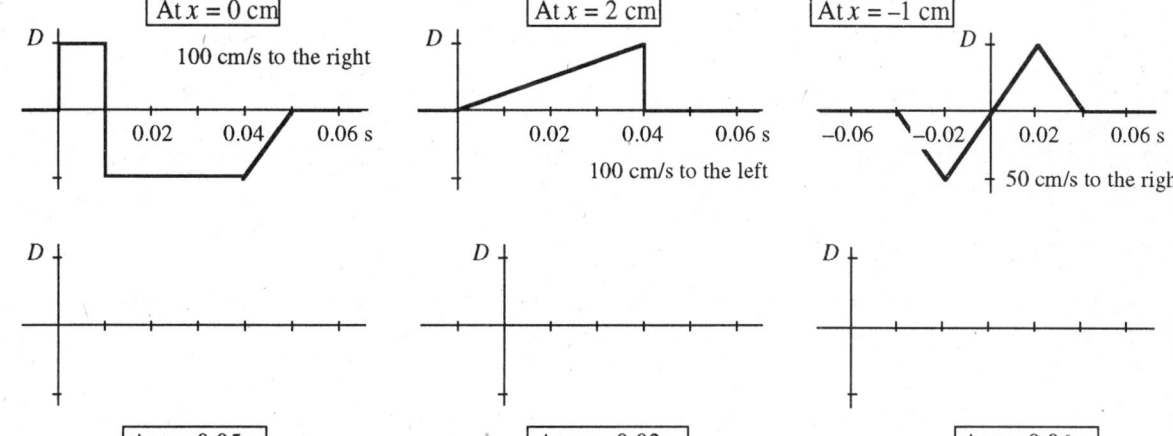

13. The figure shows a horizontal slinky at rest on a table. A wave pulse is sent along the slinky, causing the top of link #5 to move *horizontally* with the displacement-versus-time shown in the graph.

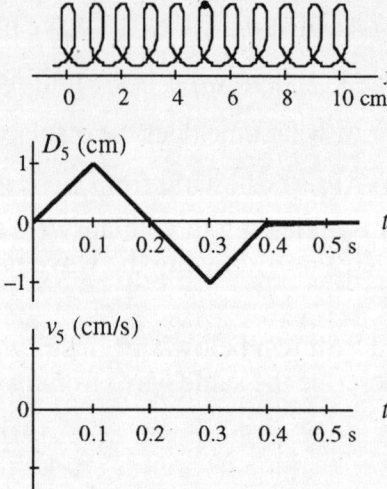

a) Is this a transverse or a longitudinal wave? Explain.

b) What is the *position* of link #5 at $t = 0.1$ s? _____

What is the position of link #5 at $t = 0.2$ s? _____

What is the position of link #5 at $t = 0.3$ s? _____

Note: This is asking for the position, *not* for the displacement.

c) Draw a velocity-versus-time graph of link #5 as the wave pulse passes. Add an appropriate scale to the vertical axis.

d) Describe *in words* how link #5 moves. That is, divide the motion into intervals — such as 0.01 s to 0.03 s — and say whether the link is moving to the right, moving to the left, or at rest. If it is moving, say what its speed is.

e) Can you determine from the information given if the wave pulse is traveling to the right or to the left? If so, give the direction and an explanation of how you found it. If not, why not?

f) Can you determine from the information given the speed of the wave? If so, give the speed and an explanation of how you found it. If not, why not?

14. Consider a long slinky on a frictionless table. You suddenly and quickly push one end of the slinky forward by 10 cm, at constant speed, then quickly stop. This push generates a compression pulse that travels down the slinky. The links in front of the pulse still have their original spacing, and those behind the pulse have a new, shorter equilibrium spacing. Notice that the total number of links in the slinky (38) is unchanged.

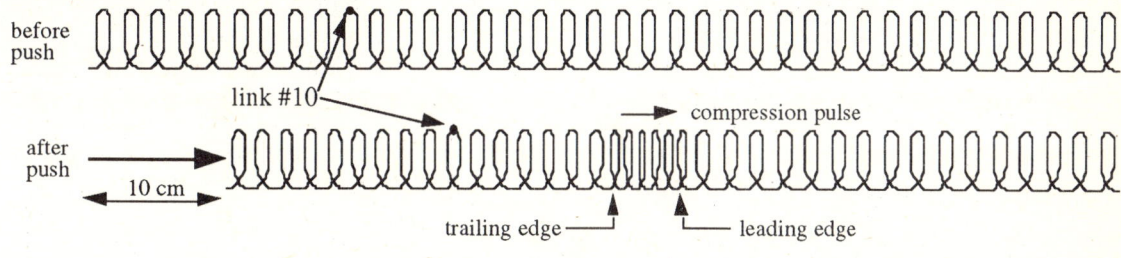

e)

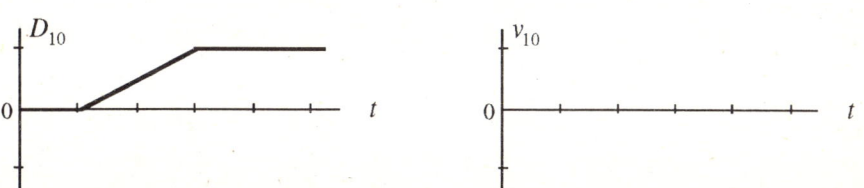

a) The left graph shows the displacement-versus-time of link #10. Draw the velocity-versus-time graph for this link on the right axes.

b) Does this link return to its starting position? _____

c) Now return the slinky to its original length. Create a new wave pulse by quickly pushing the end *forward* 10 cm, at constant speed, then immediately pulling the end *back* to its starting position. Graph the displacement and the velocity of link #10 on the axes below. Assume that the motion of link #10 begins at 1 unit of time and that it reaches its maximum displacement at 3 units of time.

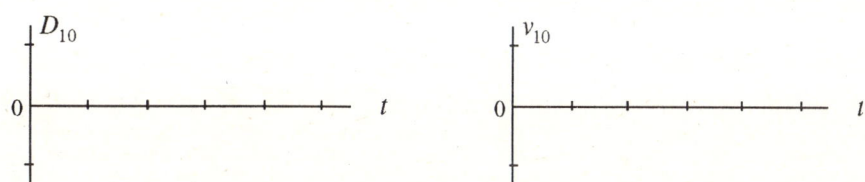

d) The links in front of the wave pulse of part c) still have their original spacing because the wave hasn't reached them. After the wave pulse has passed, the links return to their original spacing because the end of the slinky has returned to its original position. The links near the leading edge of the wave pulse are bunched together (a compression). What can you conclude about the spacing of the links near the trailing edge of the wave pulse? Explain your reasoning.

e) Draw a *picture* of the wave pulse on this slinky. Draw it in the empty space labeled e) at the top of this page, beneath the second slinky. Show the same number of links (38) as the other two slinkies. Place the leading edge of the pulse at the same location it is in the second slinky.

Chapter 14 Traveling Waves

15. Rather than draw the entire slinky, we can use a series of dots to represent the positions of the links. The first figure below shows a slinky at rest, with a 1 cm spacing between the links. A wave pulse is sent down the slinky, traveling to the right at 10 cm/s. The second picture shows the links at the instant of time $t = 0$. The links are numbered, and you can measure the displacement of each link at this instant of time.

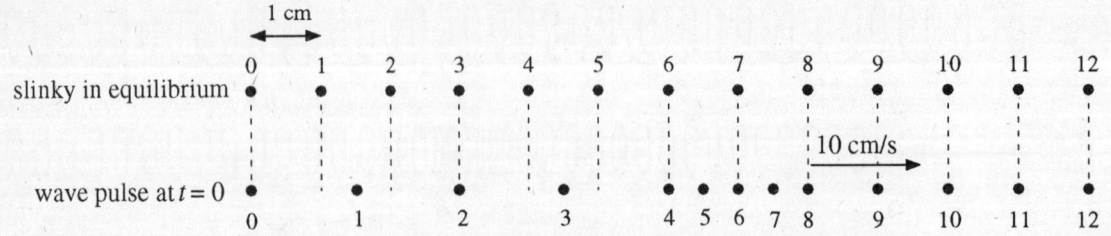

a) Draw a snapshot graph showing the displacement of each link at $t = 0$. Since there are 13 links, your graph should have 13 dots. Connect your dots with lines to make a continuous graph.

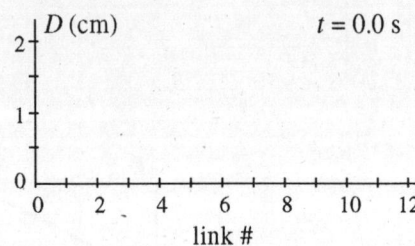

b) The slinky is horizontal. Also, the x-axis of your graph is horizontal.

Which way are the links displaced — horizontally or vertically? _____

Which way is the displacement graphed — horizontally or vertically? _____

Is your graph a "picture" of the wave or a "representation" of the wave? Explain.

c) Which links are in compression? (list their numbers) _____

Which links are in rarefaction? (list their numbers) _____

d) Draw graphs of displacement-versus-link number at $t = 0.1$ s and $t = 0.2$ s.

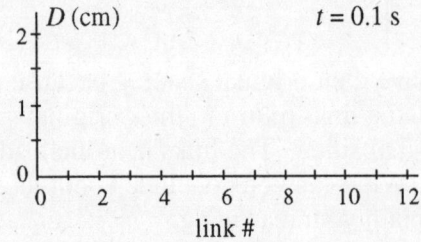 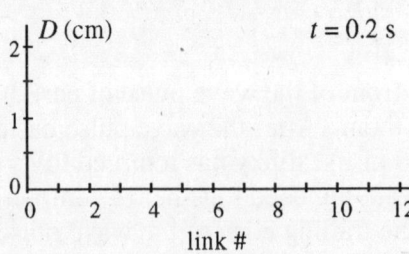

e) Now draw dot pictures of the link positions at $t = 0.1$ s and $t = 0.2$ s. The equilibrium positions are shown for reference.

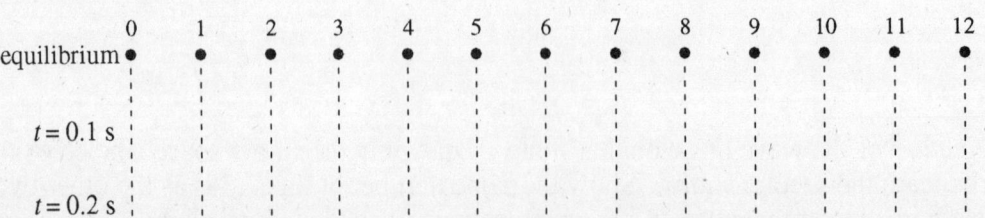

f) At $t = 0.1$, which links are moving to the right? _____

Which links are moving to the left? _____

Which links are at rest? _____

In thinking about this, you may want to compare your $t = 0$, $t = 0.1$ s, and $t = 0.2$ s pictures.

16. The graphs below show displacement-versus-link number for wave pulses on a slinky. For each, draw a dot pattern showing how the appearance of the slinky at this instant of time. A picture of the slinky at rest, with 1 cm spacings, is given for reference.

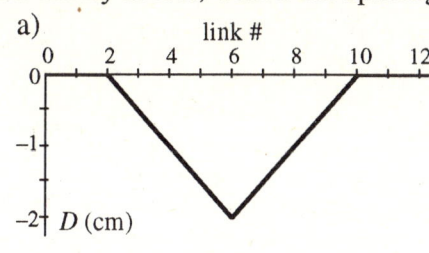
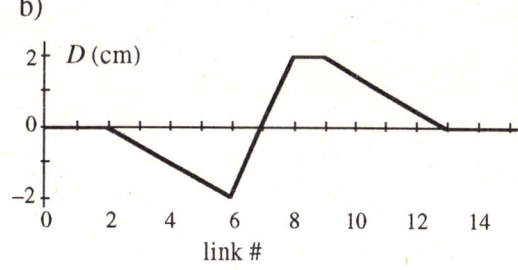
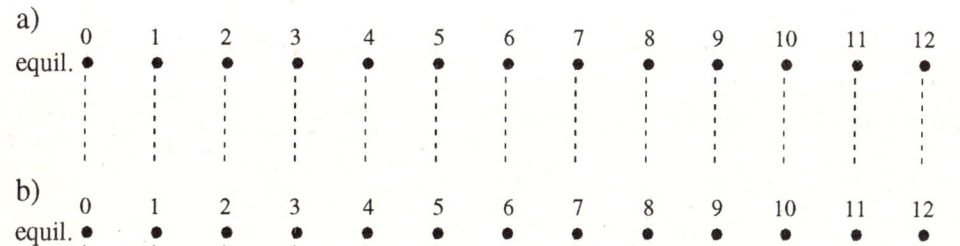

14.4 Sinusoidal Waves

14.5 Mathematics of Sinusoidal Waves

17. The figure shows a sinusoidal traveling wave. Draw a graph of the wave if:

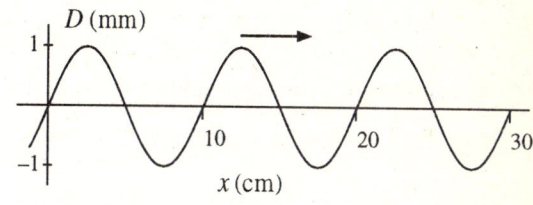

a) Its amplitude is halved and its wavelength doubled.

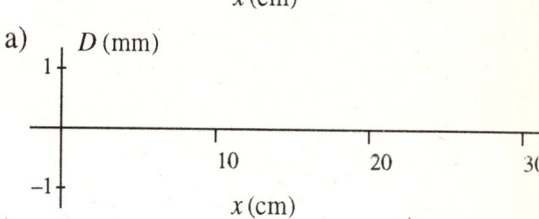

b) Its speed doubled and its frequency quadrupled.

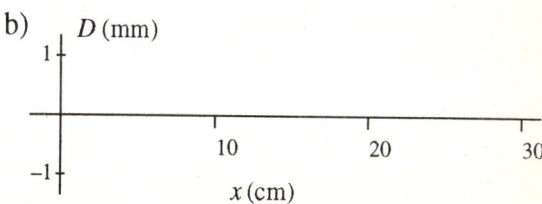

18. The wave shown at time $t = 0$ is traveling to the right at a speed of 25 cm/s.

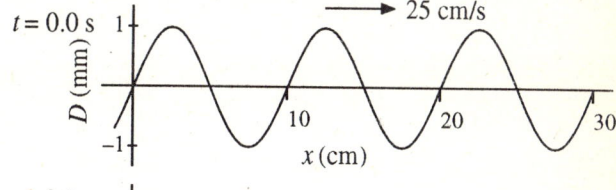

a) Draw snapshot graphs of this wave at times $t = 0.1$ s, $t = 0.2$ s, $t = 0.3$ s, $t = 0.4$ s.

b) What is the wavelength of the wave?

c) Based on your graphs, what is the period of the wave? Explain how you found it.

d) What is the frequency of the wave?

e) What is the value of the product λf?

How does this value of λf compare with the speed of the wave?

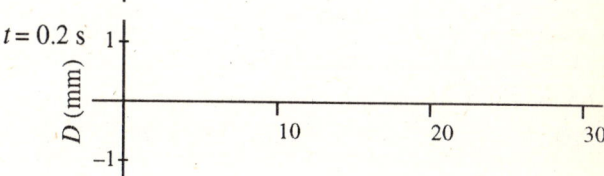
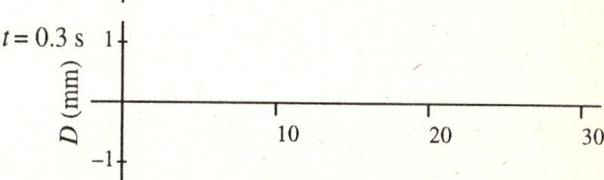
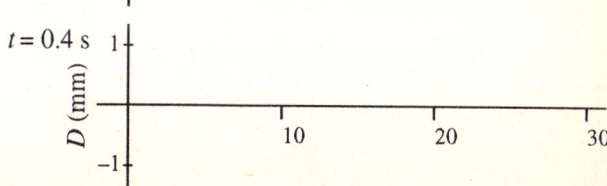

148   Chapter 14 Traveling Waves

f) How far does the wave move during one period? _____

g) In general, for any sinusoidal wave, how far does it move during a time interval of one period T?

19. A wave front diagram is shown for a sinusoidal *plane wave* at time $t = 0$. The diagram shows only the xy-plane, but keep in mind that a plane wave extends above and below the plane of the paper.

a) What is the wavelength of this wave? _____

b) At $t = 0$, for which values of y is the wave at a crest?

c) At $t = 0$, for which values of y is the wave at a trough?

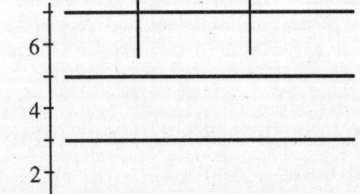

d) Can you tell from the diagram if this is a transverse or a longitudinal wave? If so, which is it and how did you determine it. If not, why not?

e) How does the displacement at the point $(x, y, z) = (6, 5, 0)$ compare to the displacement at the point $(2, 5, 0)$? Is it more, less, the same, or is there no way to tell? Explain.

f) How does the displacement at the point $(x, y, z) = (2, 5, 5)$ compare to the displacement at the point $(2, 5, 0)$? Is it more, less, the same, or is there no way to tell? Explain.

g) Draw a snapshot graph below, at $t = 0$, of displacement-versus-position for this wave. Assume that the wave's amplitude is 2 mm. Label both your axes and place an appropriate numerical scale on each axis (with units!).

h) Draw a wave front diagram at time $t = 0.3$ s.

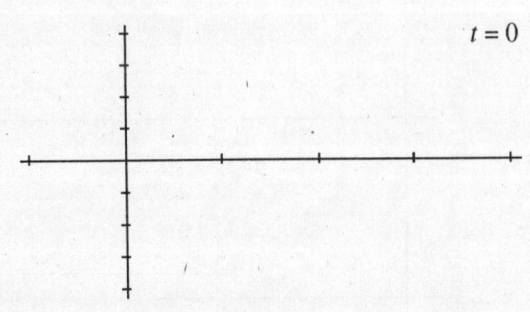
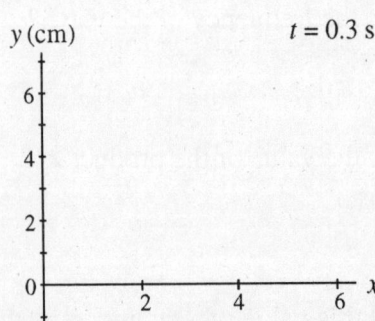

Chapter 14 Traveling Waves   149

21. a) The phase at one point on a sinusoidal wave is $(3/2)\pi$. Is the displacement at that point a crest, a trough, zero, or in between?

b) What if the phase is $(5/2)\pi$?

c) What if the phase is 5π?

22. Three waves traveling to the right are shown below. The first two are shown at $t = 0$, the third at $t = T/2$. What are the phase constants ϕ_0 for these waves?

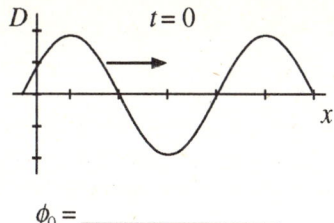 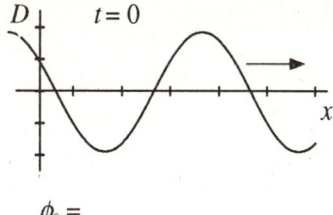 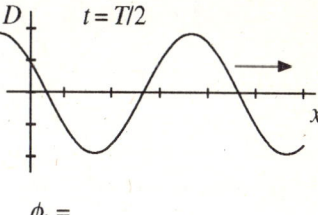

$\phi_0 = $ _____ $\phi_0 = $ _____ $\phi_0 = $ _____

Note: Knowing the displacement at a particular point in space and time, such as $D(0,0)$ is a *necessary* piece of information for finding ϕ_0 but is not by itself enough. The first two waves above have the same value for $D(0,0)$ but they do *not* have the same ϕ_0. You must also consider the overall shape of the wave.

23. A plane wave of wavelength 2 m is traveling along the x-axis. At $t = 0$ the wave's phase at $x = 2$ m on the axis $(x, y, z) = (2, 0, 0)$ m is $\pi/2$.

a) Draw a snapshot graph of the wave at $t = 0$.

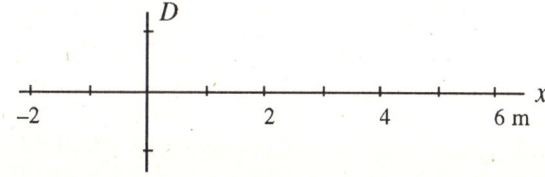

b) At $t = 0$, what is the phase at the point $(0, 0, 0)$ m? _____

c) At $t = 0$, what is the phase at the point $(1, 0, 0)$ m? _____

d) At $t = 0$, what is the phase at the point $(3, 0, 0)$ m? _____

e) At $t = 0$, what is the phase at the point $(2, 2, 2)$ m? _____

Note: No calculations are needed to answer these. You just need to think about what the phase *means* and utilize your graph from a).

24. Consider the wave shown. Redraw this wave on the axes on the next page if

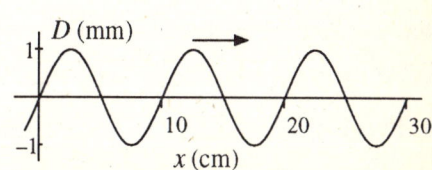

a) Its wave number is doubled.

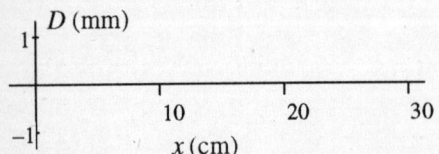

b) Its wave number is halved.

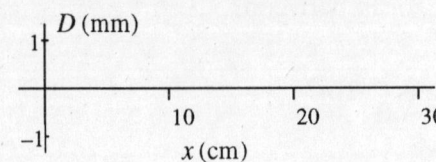

14.6 Sound and Light

14.7 Power and Intensity

14.8 The Principle of Superposition

25. The two waves arrive simultaneously at a point in space from two different sources.

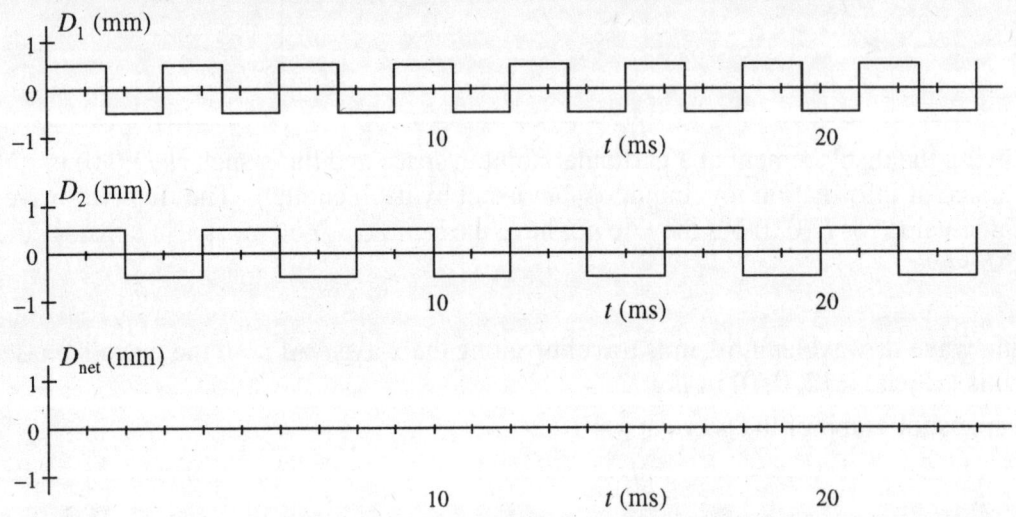

a) Period of wave #1? _____   b) Period of wave #2? _____

Frequency of wave #1? _____   Frequency of wave #2? _____

c) Draw the graph of the net wave at this point on the axes above. Be accurate, use a ruler!

d) Period of the net wave? _____   Frequency of the net wave? _____

e) Is the frequency of the superposition what you would expect as a beat frequency? Explain.

26. The figure shows a snapshot graph of two plane waves passing through a region of space. Each has a 2 mm amplitude. For each of the lettered points, determine the displacements of the individual waves (numbered 1 and 2) and the net displacement.

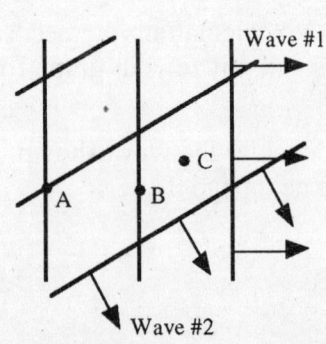

a) Point A: $D_1 =$ _____ $D_2 =$ _____ $D_{net} =$ _____

b) Point B: $D_1 =$ _____ $D_2 =$ _____ $D_{net} =$ _____

c) Point C: $D_1 =$ _____ $D_2 =$ _____ $D_{net} =$ _____

Chapter 15

Standing Waves

15.1 Waves with Boundaries

15.2 Reflections and Superposition

1. The two graphs below are snapshot graphs at $t = 0$ of a wave pulse on a string as it approaches a boundary. It is traveling to the right with a speed of 1000 cm/s. Draw snapshot graphs of the wave pulse at $t = 4$ ms, $t = 12$ ms, and $t = 20$ ms. Don't forget that the displacement reverses upon reflection.

a) 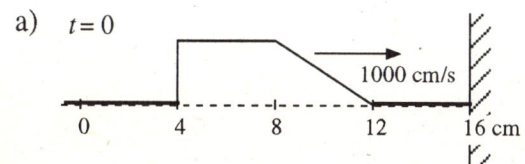 b)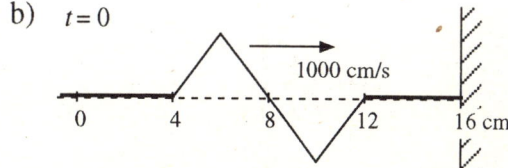

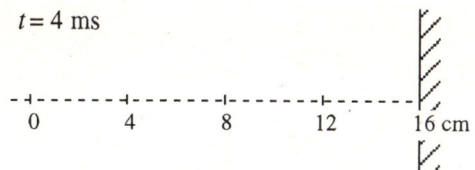 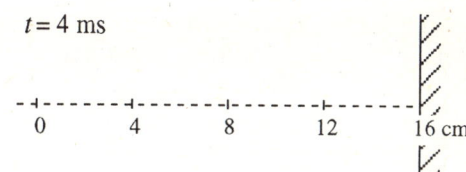

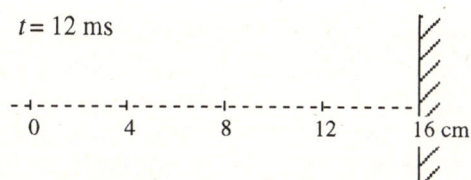 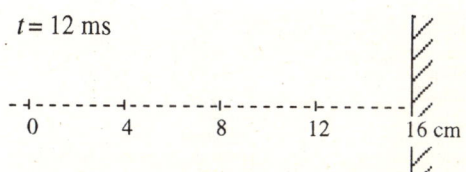

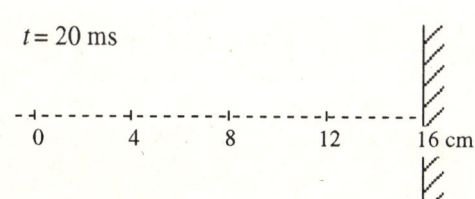 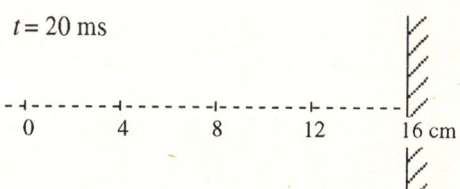

152 Chapter 15 Standing Waves

2. Look more closely at how the wave pulse of Exercise 1b) reflects. The first graph below shows the wave pulse, traveling to the right, at $t = 4$ ms. Consider the situation every millisecond: $t = 4$ ms, 5 ms, 6 ms, ..., 12 ms.

a) On the left axes below, draw graphs at the indicated value of t for just that portion of the wave that is traveling *to the right*. As a piece of the wave reaches the boundary, it disappears from these graphs. Use a ruler so that your graph is sufficiently accurate to perform step c).

b) Then on the middle set of axes, for the same values of t, draw graphs of the reflecting wave as it travels *to the left*. There is no reflecting wave at $t = 0$, but it will emerge frame-by-frame as you proceed with your graphs. Don't forget the displacement reversal upon reflection.

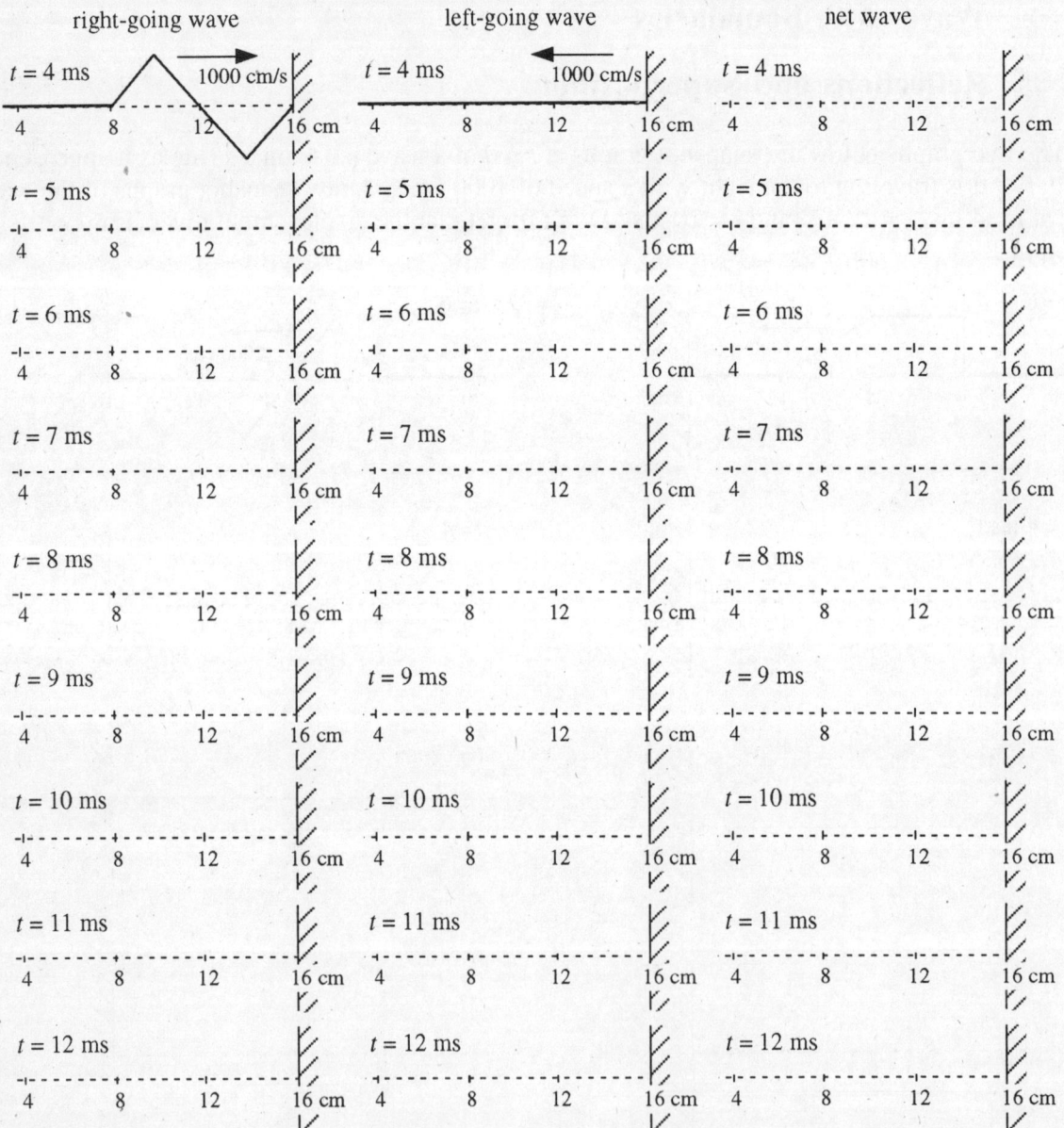

c) Finally, apply the principle of superposition to find the net displacement of the string at each of these instants of time. Draw these graphs on the right axes. Use a ruler to be accurate.

Chapter 15 Standing Waves

15.3 Standing Waves on a String

3. Two waves are traveling along a string, one to the right and one to the left. Each has a speed of 1000 cm/s and an amplitude of 1 cm. The first set of graphs below shows each wave at $t = 0$ as it would appear *if* the other wave did not exist.

a) Draw the superposition of these two waves at $t = 0$ on the axes to the right. This superposition shows the net displacement of the string, D_{net}, at time $t = 0$.

b) On the axes at the left, draw each of the two waves separately every 1 ms until $t = 8$ ms. Don't forget that the waves extend beyond the graph edges, so new pieces of the wave will move in to take the place of pieces that move away from the edges.

c) On the axes at the right, draw the superposition of the two waves at these same values of t.

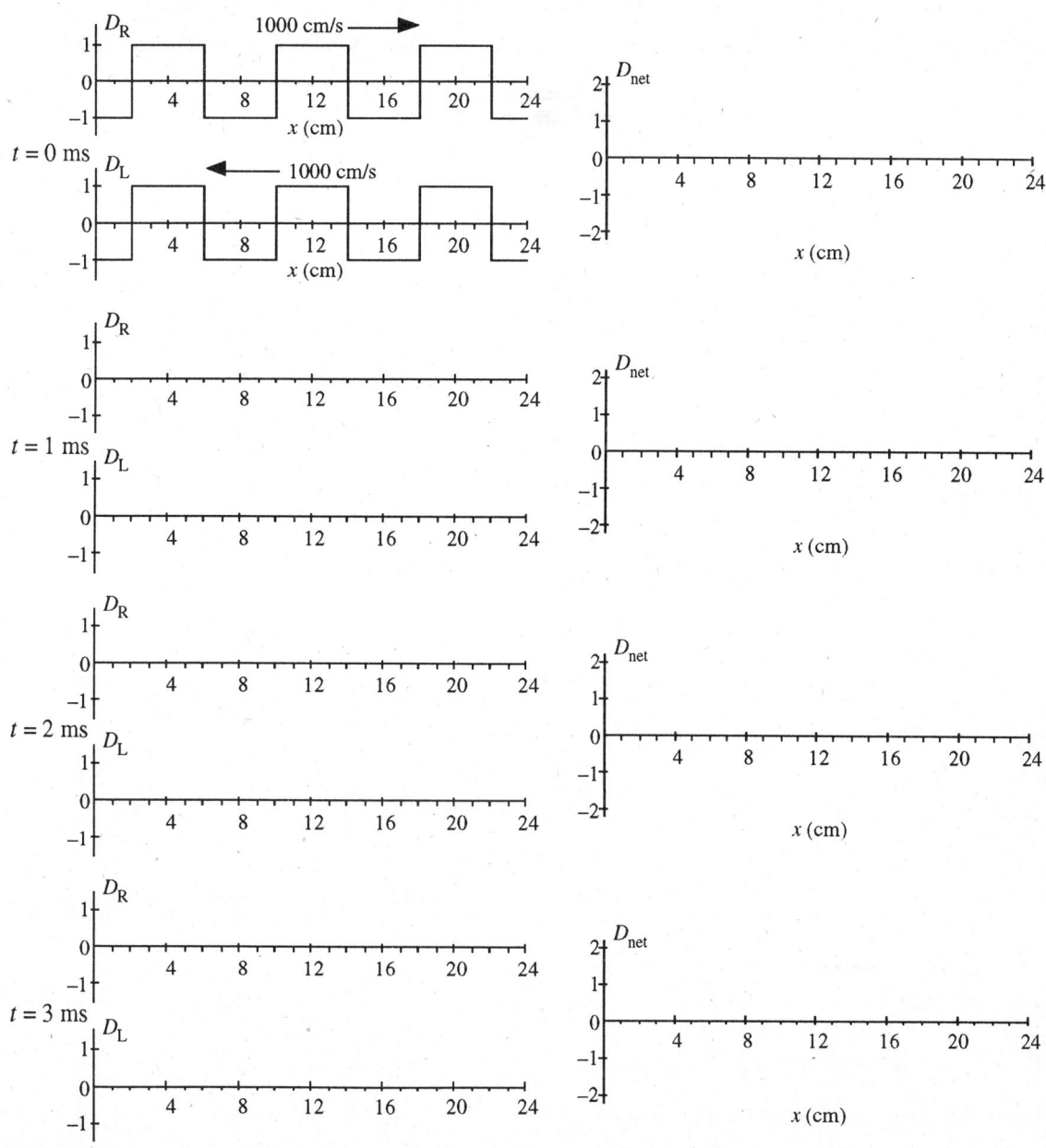

Continues next page:

154   Chapter 15 Standing Waves

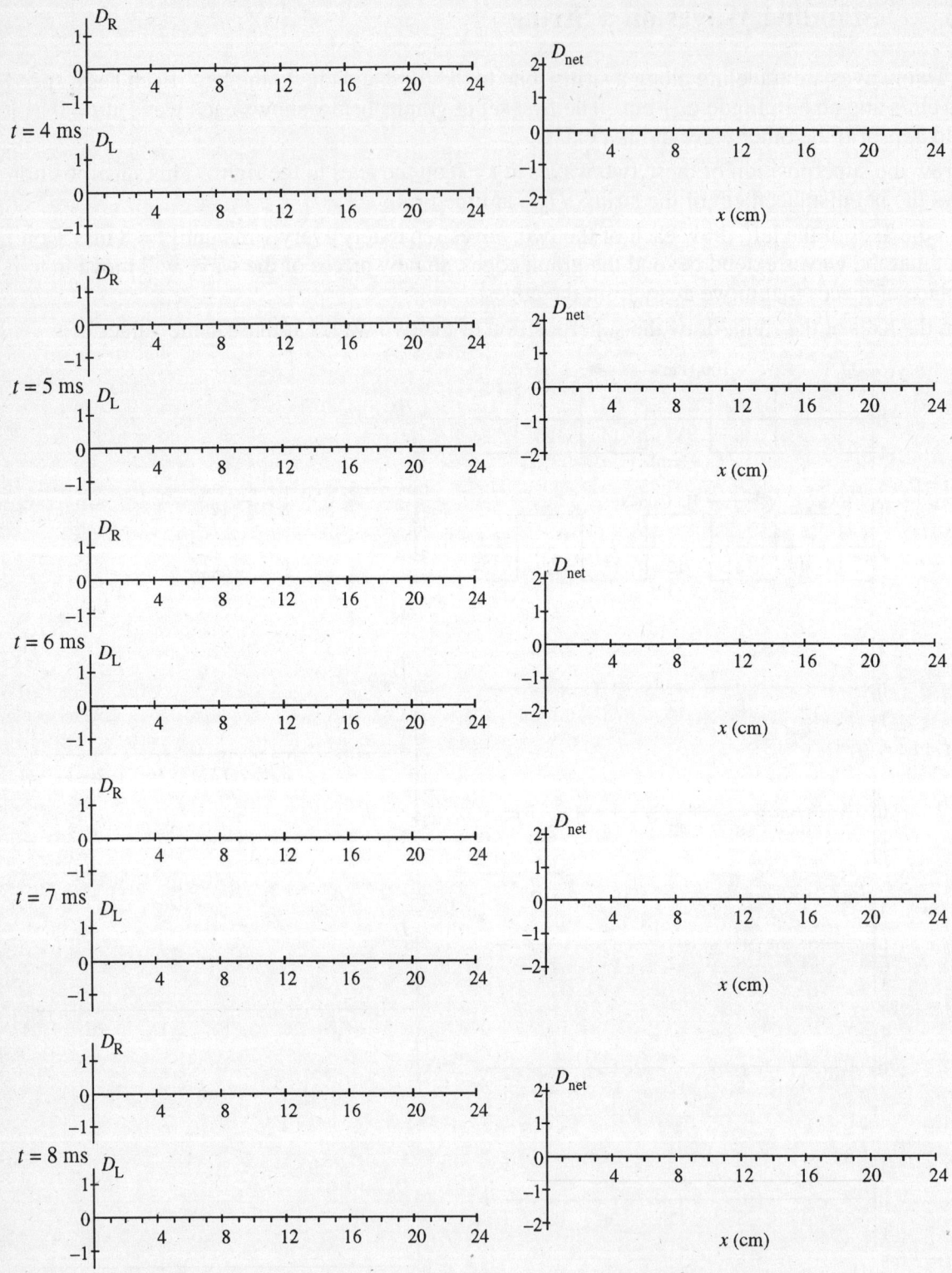

d) What is the wavelength of the traveling waves? _____ Of the net wave? _____

What is the period of each traveling wave? _____ Of the net wave? _____

f) Are their times when the string has <u>no</u> displacement at any point? When? _____

4. a) This string has two triangular waves traveling to the right and to the left at 1000 cm/s, each with a 1 cm amplitude. Repeat the same a), b), and c) steps of Exercise 3. You'll need to think carefully about adding the waves at some instants of time. Use a ruler.

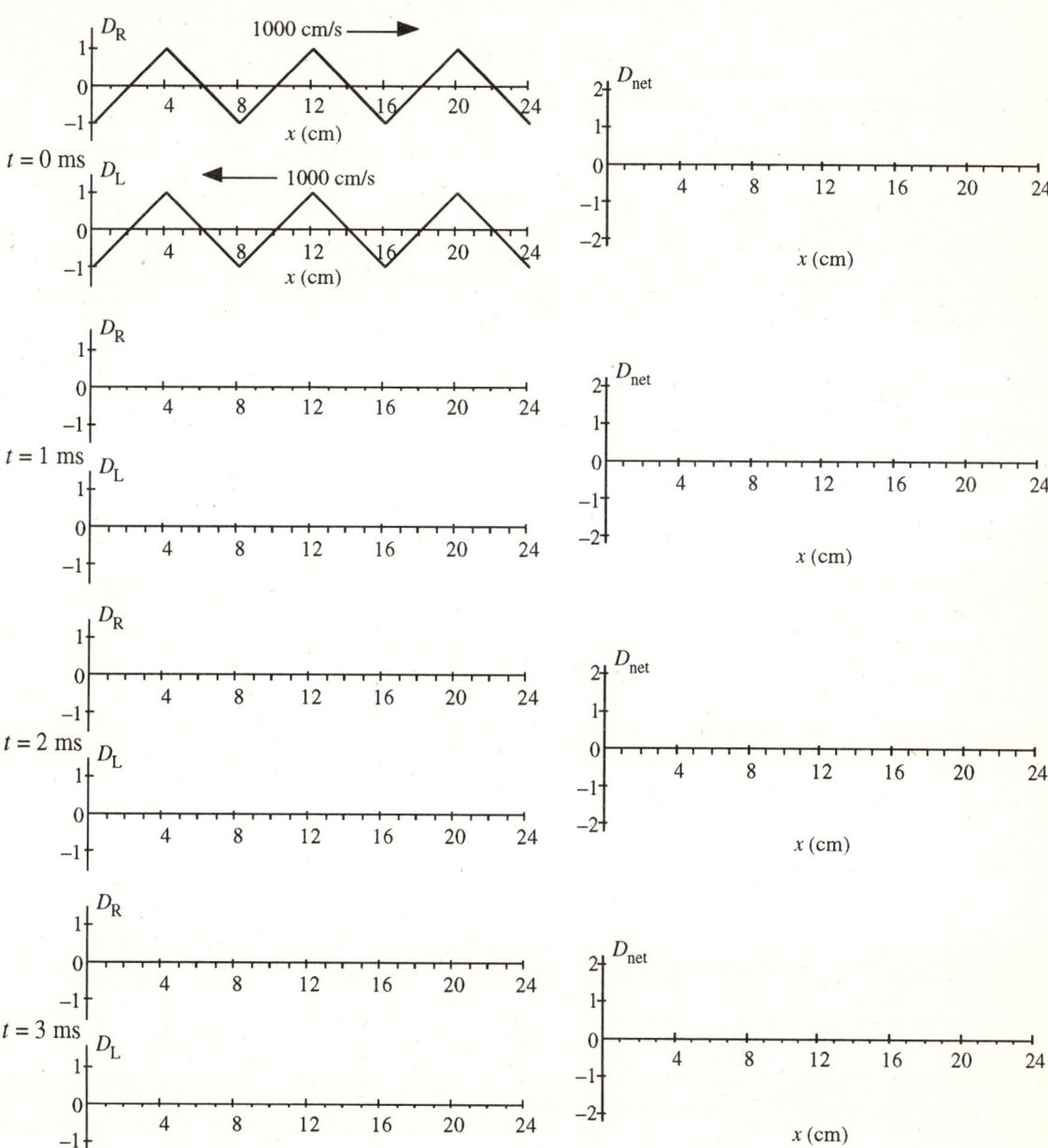

Continues next page:

156 Chapter 15 Standing Waves

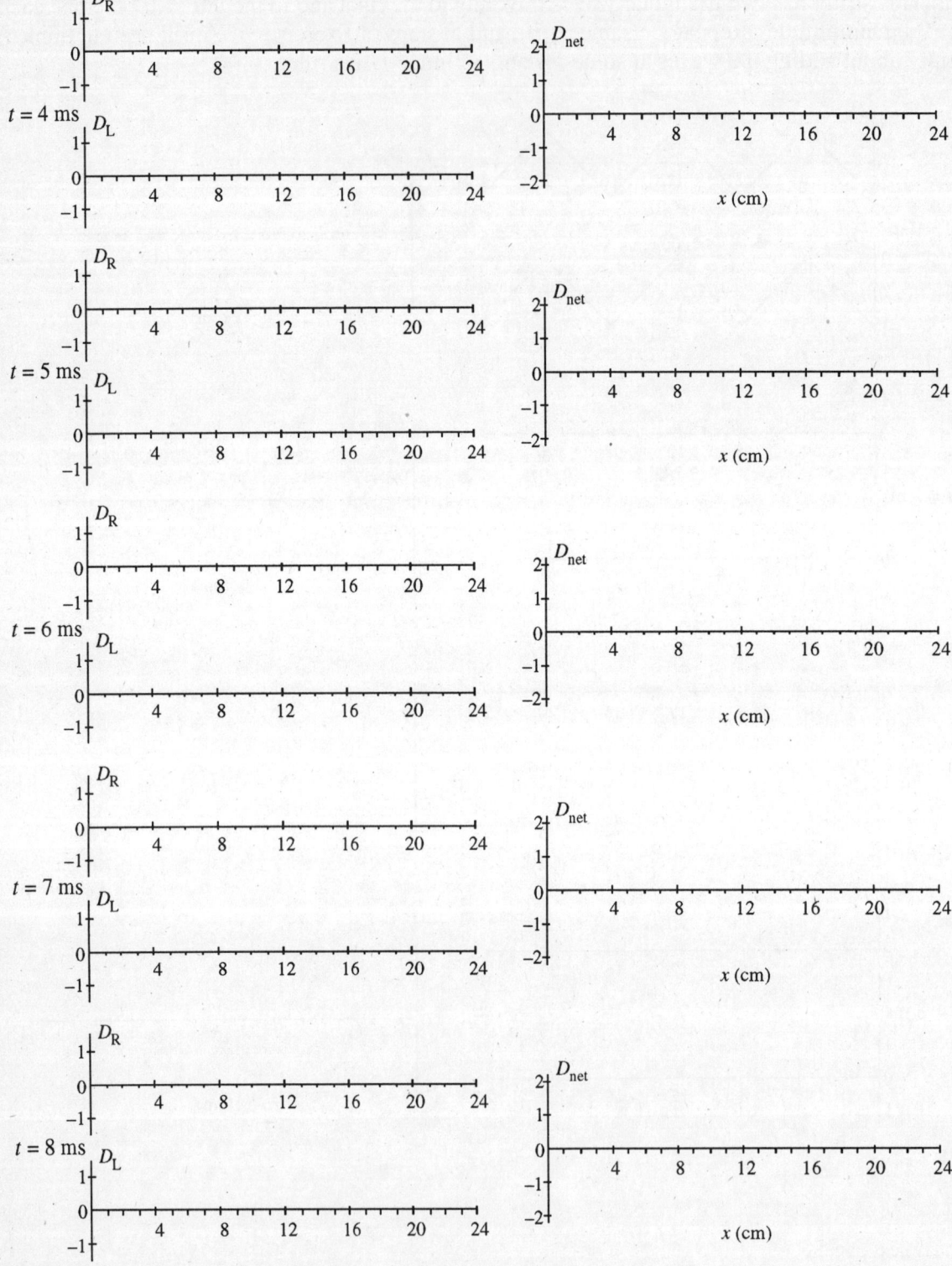

b) What is the wavelength of the standing wave? _____

c) Are there points on the string that *never* move? Where? _____

How far apart are each of these points? _____

What fraction of a wavelength is the distance between these points? _____

d) Are there points on the string that oscillate with a larger amplitude than other points?

Where? _____ How far apart are each of these points? _____

d) Does the net wave appear to *travel* to the right or the left? If so, which way?

5. The first graph shows a wave on a string at $t = 0$. This wave has a period of 8 ms. Draw snapshot graphs of the string every 1 ms from $t = 1$ ms to $t = 8$ ms. Think carefully about the proper amplitude at each instant of time.

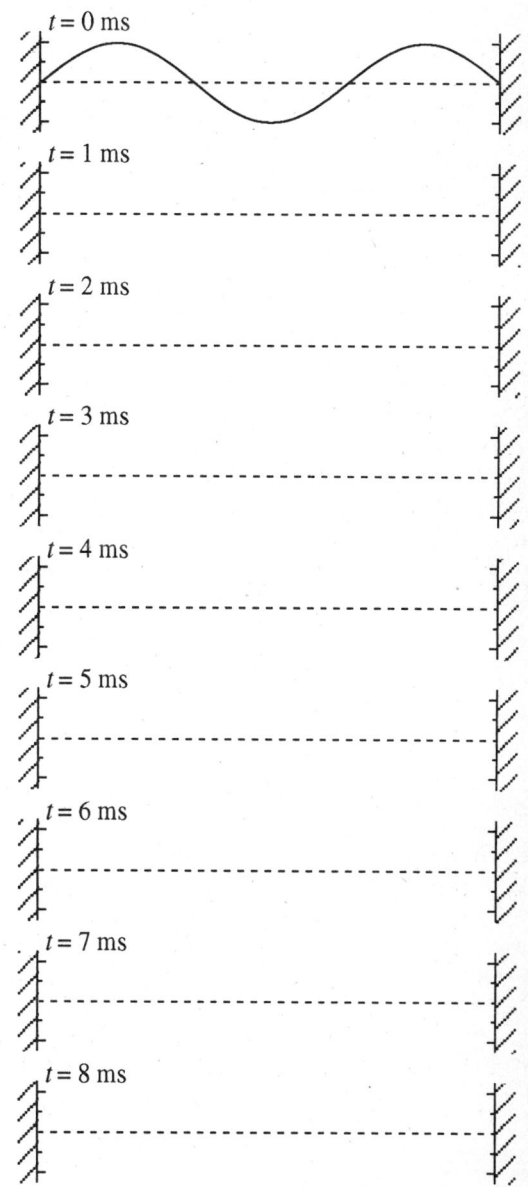

15.4 The Wavelengths and Frequencies of Standing Waves

6. The figure shows a standing wave on a string. It has frequency f.
a) Draw the standing wave that occurs if the frequency is changed to $(2/3)f$.
b) Draw the standing wave that occurs if the frequency is changed to $(3/2)f$.

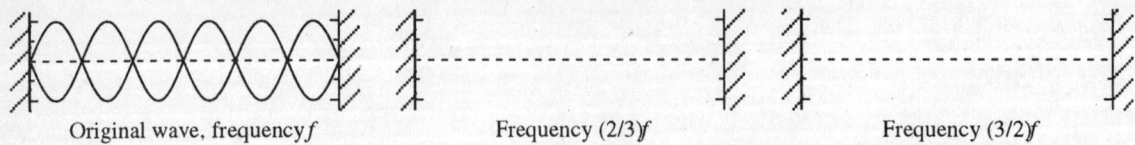

c) Is there a standing wave if the frequency is changed to $(1/4)f$? If so, how many antinodes will it have? If not, why not?

7. The figure shows a standing wave on a string.
a) Draw the standing wave that occurs if the string tension is quadrupled while the frequency is held constant.

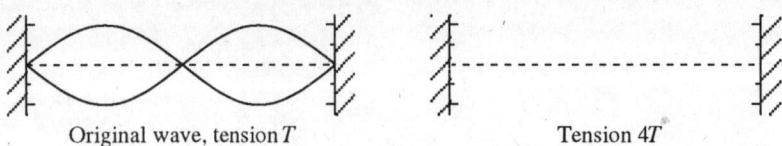

b) Suppose the tension is merely doubled while the frequency shaking the string remains constant. Will there be a standing wave? If so, how many antinodes will it have? If not, why not?

15.5 Standing Electromagnetic Waves

No exercises.

15.6 Standing Sound Waves

8. The picture shows a standing sound wave in a 32 mm long tube of air that is open at both ends.

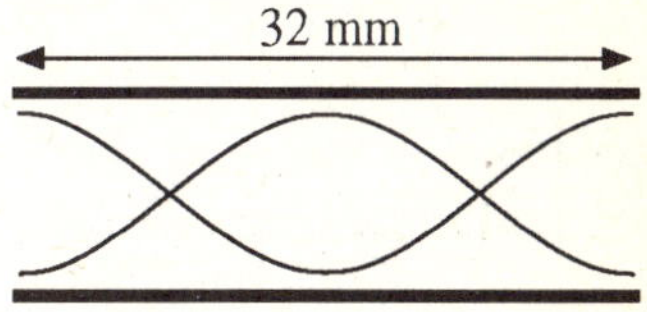

a) Which mode (value of n) standing wave is this? _____

b) Are the air molecules vibrating vertically or horizontally? Explain.

c) The purpose of this exercise is to help you visualize the motion of the air molecules that this picture represents. On the next page are a series of nine graphs, made every one-eighth of a period from $t = 0$ to $t = T$. Each graph represents the displacements of the molecules in a 32 mm long tube at that instant of time. Positive values are displacements to the right, negative values are displacements to the left. The entire series of graphs, superimposed, make the standing wave shown above.

Consider 9 air molecules that, in equilibrium, are 4 mm apart and lie along the axis of the tube. The top picture on the right shows these molecules in their equilibrium positions. The dotted lines down the page — spaced 4 mm apart — are reference lines showing the equilibrium positions. Read each graph carefully, then draw 9 dots to show the positions of the 9 air molecules at each instant of time. The first one, for $t = 0$, has already been done for you to illustrate the procedure.

(Note: For drawing purposes, it's a good approximation to assume that the first dot moves in the pattern 4 - 3 - 0 - –3 - –4 - –3 - 0 - 3 - 4 mm, the second dot in the pattern 3 - 2 - 0 - –2 - –3 - –2 - 0 - 2 - 3 mm, and so on. Most people find it easiest to trace one dot through all 9 times, then the next dot, and so on.)

When you are done, you have a good picture of the air "sloshing" back and forth inside the tube.

d) At what times is the air farthest to the right? _____   Farthest to the left? _____

   What is the difference between these times, as a fraction of a period? _____

e) At what times are *all* the air molecules in their equilibrium positions? _____

   What is the difference between these times, as a fraction of a period? _____

f) At what times does the air reach maximum compression, and where does it occur?

   Max compression at time _____   Max compression at position _____

   _____   _____

   _____   _____

g) What is the relationship between the positions of maximum compression and the nodes of the standing wave?

f) Can you determine, from the information provided, what the diameter of the tube is? If so, what is it? If not, why not?

15.7 Standing Wave Resonances

No exercises.

160 Chapter 15 Standing Waves

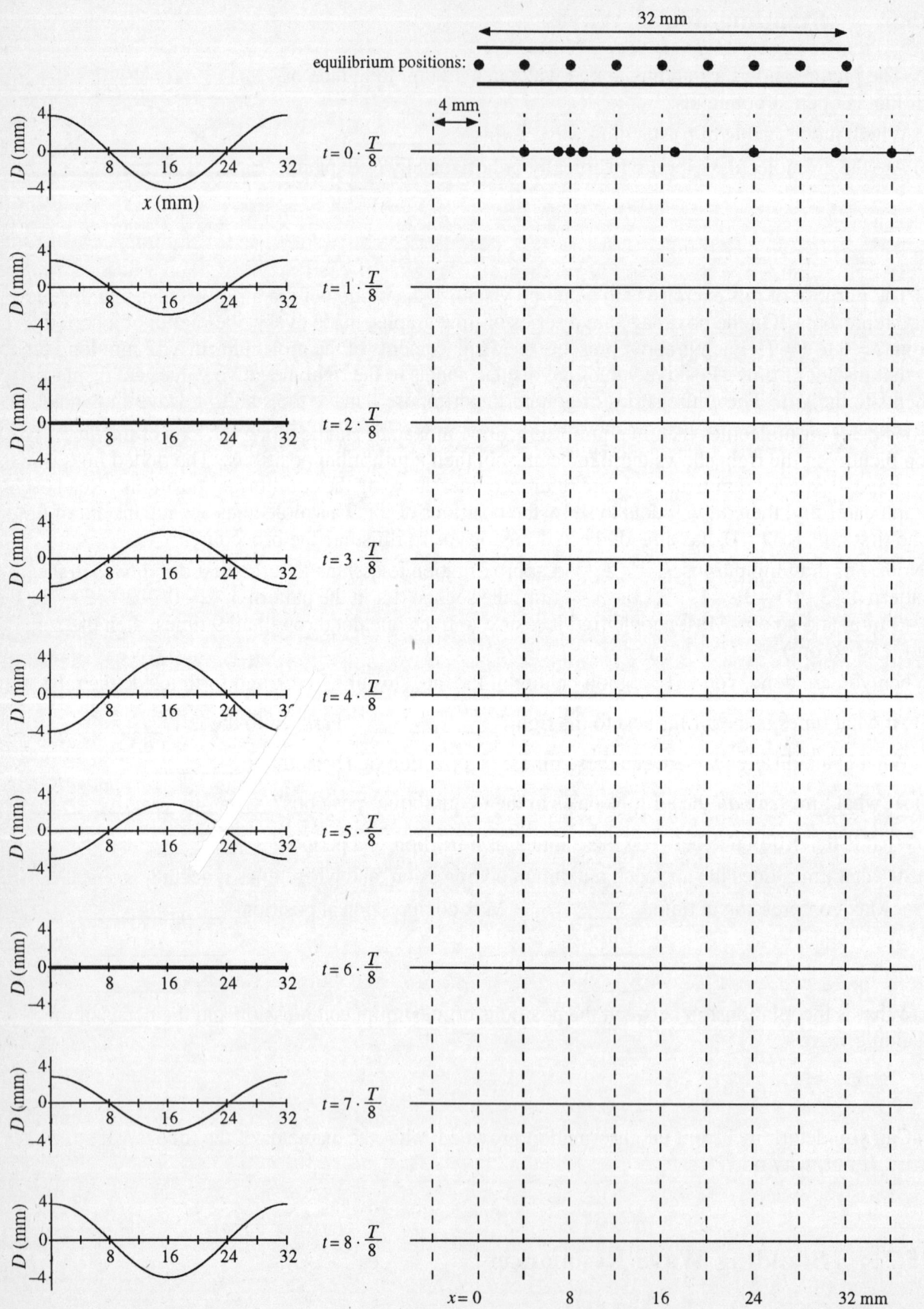

Chapter 16

Interference

16.1 Wave Interactions

16.2 Interference in One Dimension

1. The figure shows a loudspeaker emitting a triangular-shaped sound wave. This is a snapshot graph at $t = 0$. A second loudspeaker, which can be moved forward or backward, also sends a sound wave along the same axis. Both speaker cones vibrate in phase at the same frequency. The second speaker is drawn below the first, so that the figure is clear, but you want to think of it as *in front of* or *behind* the first so that the two waves are overlapped as they travel down the x-axis.

a) On the left set of axes, draw the $t = 0$ snapshot graph of the second wave if speaker #2 is placed at each of the positions shown. The first graph, with $x_{speaker} = 2$ m, is already drawn.

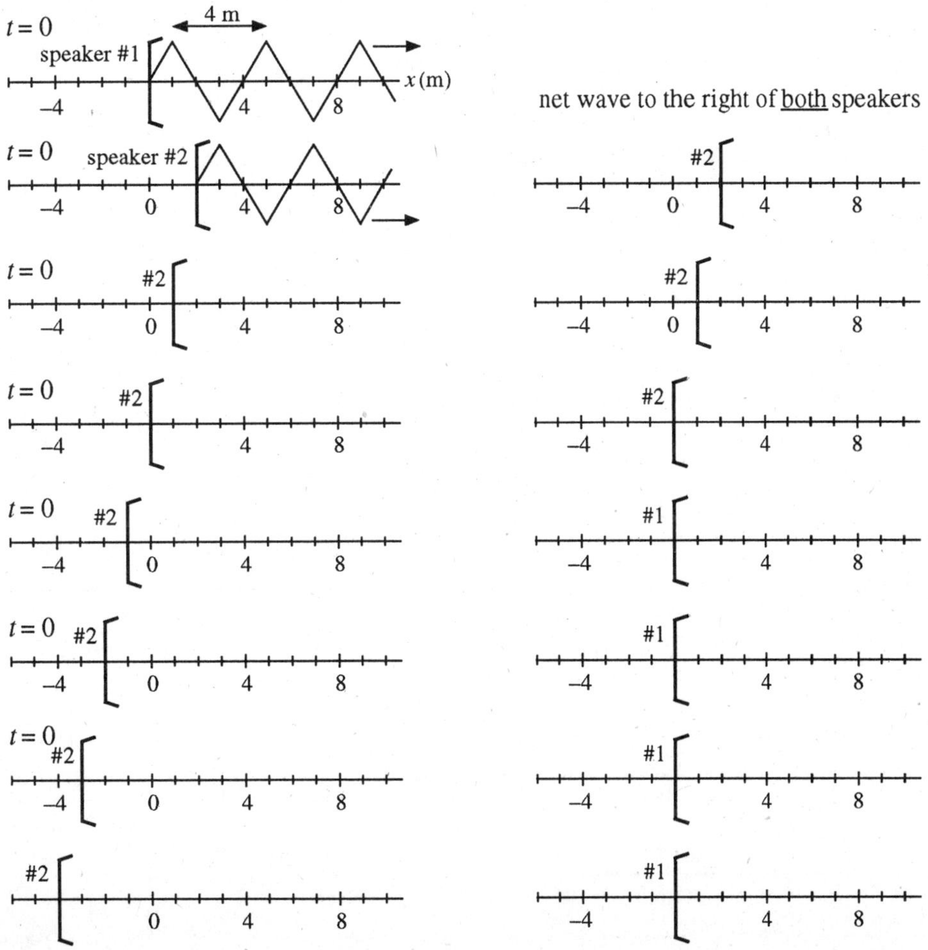

162 Chapter 16 Interference

b) On the right set of axes, draw the superposition $D_{net} = D_1 + D_2$ of the waves from the two speakers. D_{net} exists only to the right of *both* speakers and is the net wave traveling to the right.

c) What is the wavelength of these two waves? _____

d) What separations between the speakers gives constructive interference? _____

For each of these, what is the *ratio* (separation distance/λ)?

e) What separations between the speakers gives destructive interference? _____

For each of these, what is the *ratio* (separation distance/λ)?

2. Consider the same two loudspeakers as in Exercise 1.

a) Speaker #2 is 2 m behind Speaker 1. Copy your two $t = 0$ snapshot graphs from Exercise 1 onto the first set of axes. Then draw the $t = 0$ superposition graph D_{net} on the right. This simply repeats your fifth set of graphs from Exercise 1.

b) Now draw snapshot graphs of the two waves on the left, and their superposition on the right, at times $t = (1/4)T$, $(2/4)T$, and $(3/4)T$ where T is the wave's period. The waves from the two speakers will move $\lambda/4$ to the right during each of these intervals.

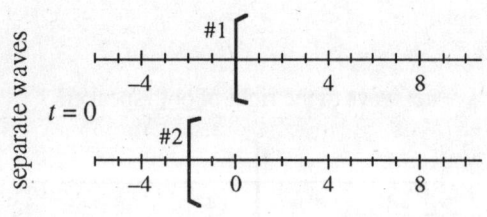
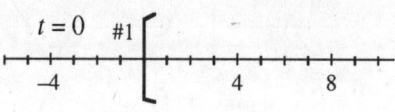

net wave to the right of both speakers

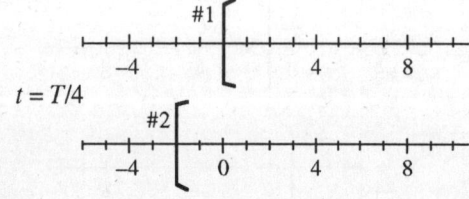
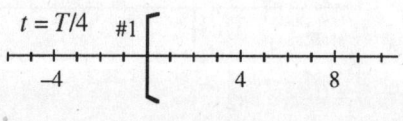

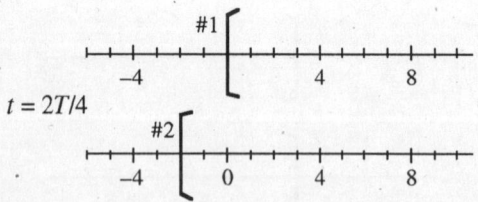
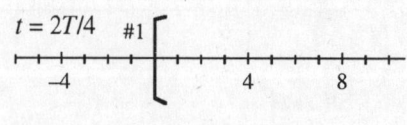

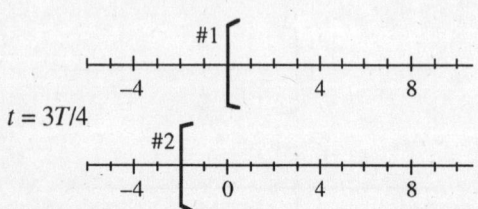

c) Repeat steps a) and b) for the same four instants of time if speaker #2 is placed 4 m behind speaker #1.

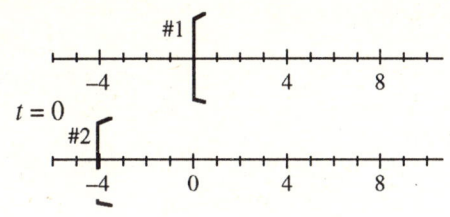
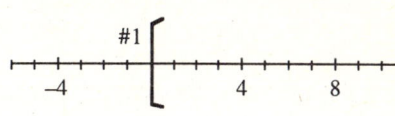

net wave to the right of both speakers

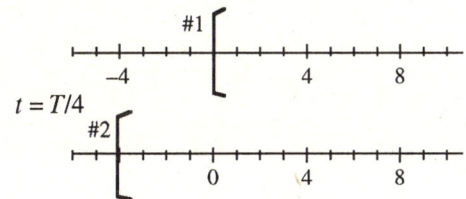
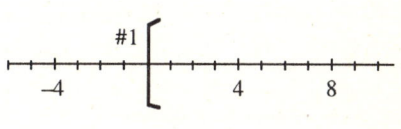

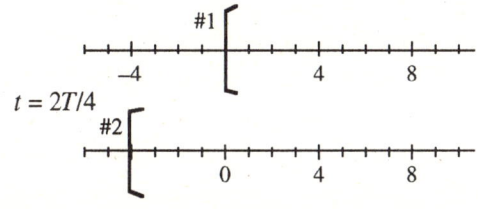
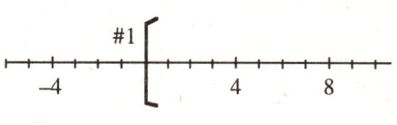

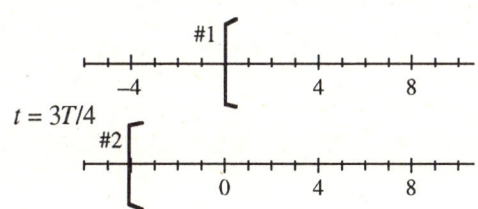
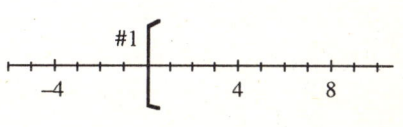

d) Speaker #2 is still 4 m behind speaker #1. Draw the *history* graph $D_{net}(x = +4 \text{ m}, t)$ showing the net displacement of the wave at the point $x = +4$ m for the interval $0 \le t \le 2T$.

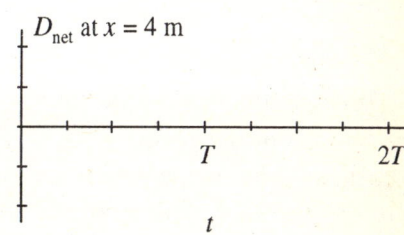

e) Is the net wave a traveling wave or a standing wave. Explain, based on what you have observed in this exercise.

164   Chapter 16 Interference

3. The phase of a wave at a point on the axis is $\phi = 2\pi(x/\lambda - t/T) + \phi_0$ where x is the *distance* from the point to the wave source (not necessarily the same value as the *position*) and ϕ_0 is the phase constant of the wave source.

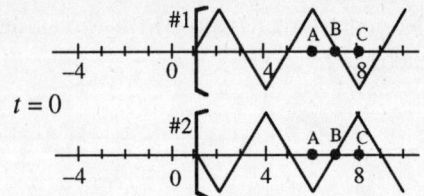
$t = 0$

a) What is the phase constant ϕ_0 of wave #1? _____

b) Determine the phase of the wave #1 at time $t = 0$ at the points on the axis marked A, B, and C. First determine the distance x, then compute the phase. Write the phase as a multiple of π. Point A is already done as an illustration.

	x	ϕ
Point A	5 m	2.5π
Point B	___	___
Point C	___	___

c) What is the phase constant of wave #2? _____

d) Determine the phase of wave #2 at time $t = 0$ at the points A, B, and C.

	x	ϕ
Point A	___	___
Point B	___	___
Point C	___	___

4. Two loudspeakers are shown at $t = 0$, with speaker #2 placed 4 meters behind speaker #1.

a) Is the interference constructive or destructive?

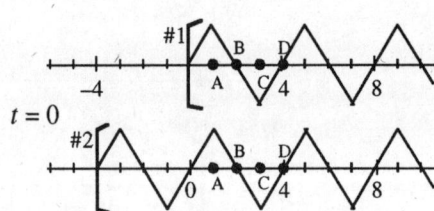
$t = 0$

b) At a point on the x-axis, the *phase* of waves #1 and #2 are $\phi_1 = 2\pi(x_1/\lambda - t/T) + \phi_{10}$ and $\phi_2 = 2\pi(x_2/\lambda - t/T) + \phi_{20}$. Here x_1 is the distance from the point to speaker #1 and x_2 is the distance to speaker #2. Note that $x_1 \neq x_2$.

What is the phase constant ϕ_{10} for wave #1? _____ ϕ_{20} for wave #2? _____

c) At each of the points A, B, C, and D on the x-axis: what are the distances x_1 and x_2 to the two speakers, what is the difference $\Delta x = |x_2 - x_1|$ between the travel distances of the two waves, what are the phases ϕ_1 and ϕ_2 of the two waves (at $t = 0$), and what is the phase difference $\Delta\phi = |\phi_2 - \phi_1|$ between the waves? Point A is already filled in to illustrate.

	x_1	x_2	Δx	ϕ_1	ϕ_2	$\Delta\phi$
Point A	1 m	5 m	4 m	0.5π	2.5π	2π
Point B	___	___	___	___	___	___
Point C	___	___	___	___	___	___
Point D	___	___	___	___	___	___

d) Now speaker #2 is placed only 2 m behind speaker #1. Is the interference constructive or destructive?

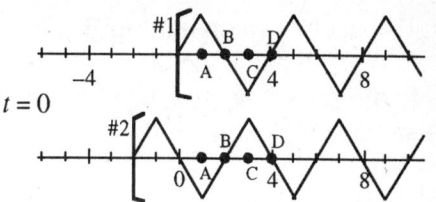

e) At each of the points A, B, C, and D on the x-axis: what are the distances x_1 and x_2 to the two speakers, what is the difference $\Delta x = |x_2 - x_1|$ between the travel distances of the two waves, what are the phases ϕ_1 and ϕ_2 of the two waves (at $t = 0$), and what is the phase difference $\Delta\phi = |\phi_2 - \phi_1|$ between the waves?

	x_1	x_2	Δx	ϕ_1	ϕ_2	$\Delta\phi$
Point A	___	___	___	___	___	___
Point B	___	___	___	___	___	___
Point C	___	___	___	___	___	___
Point D	___	___	___	___	___	___

5. Two speakers are placed side-by-side at $x = 0$, but now the wave emitted by speaker #2 has been changed.

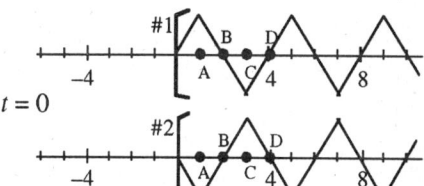

a) What is the wavelength of wave #1? _____

    Of wave #2? _____

b) What is the phase constant ϕ_{10} for wave #1? _____

    What is ϕ_{20} for wave #2? _____

c) Is the interference of these two waves constructive or destructive? _____

d) At each of the points A, B, C, and D on the x-axis: what are the distances x_1 and x_2 to the two speakers, what is the difference $\Delta x = |x_2 - x_1|$ between the travel distances of the two waves, what are the phases ϕ_1 and ϕ_2 of the two waves, and what is the phase difference $\Delta\phi = |\phi_2 - \phi_1|$ between the waves?

	x_1	x_2	Δx	ϕ_1	ϕ_2	$\Delta\phi$
Point A	___	___	___	___	___	___
Point B	___	___	___	___	___	___
Point C	___	___	___	___	___	___
Point D	___	___	___	___	___	___

e) Speaker #2 is moved back 2 m. Does this change its phase constant ϕ_0?

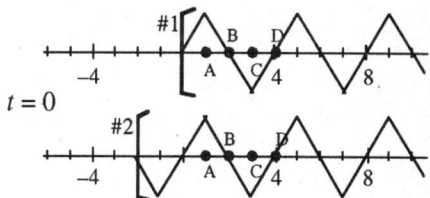

f) Is the interference constructive or destructive?

166   Chapter 16 Interference

g) At each of the points A, B, C, and D on the x-axis: what are the distances x_1 and x_2 to the two speakers, what is the difference $\Delta x = |x_2 - x_1|$ between the travel distances of the two waves, what are the phases ϕ_1 and ϕ_2 of the two waves, and what is the phase difference $\Delta\phi = |\phi_2 - \phi_1|$ between the waves?

	x_1	x_2	Δx	ϕ_1	ϕ_2	$\Delta\phi$
Point A						
Point B						
Point C						
Point D						

6. Review your answers to the first five exercises. Is it the separation Δx between the speakers or the phase difference $\Delta\phi$ between the waves that determines whether the interference is constructive or destructive? Explain.

7. The two speakers are shown at $t = 0$. The distance between tick marks for speaker #2 is the same 1 m as for speaker #1, but the location of the origin is unknown.

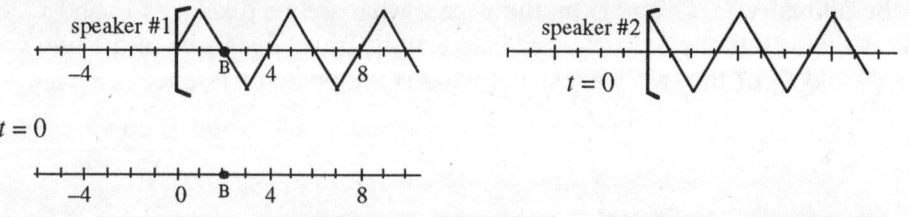

a) What is the phase constant ϕ_{10} for speaker #1? _____ ϕ_{20} for speaker #2? _____
b) Place speaker #2 on the axis behind speaker #1 so that it will interfere constructively with speaker #1. Do this by drawing speaker #2 on the empty axes below speaker #1. Then draw wave #2.
c) What is Δx, the difference between the distances to the speakers, at point B? _____
d) What is $\Delta\phi$, the phase difference between the two waves, at point B? _____

16.3 Interference in Two and Three Dimensions

8. Speakers #1 and #2 are 12 m apart. Both are emitting identical triangular sound waves with $\lambda = 4$ m and $\phi_0 = 0$. Point A is distance $r_1 = 16$ m from speaker #1. The wave fronts show the crests of wave #1 at time $t = 0$. They are numbered 1 to 5. (The wave fronts actually should be circular arcs, but it is easier to use straight lines to show where the wave fronts cross the axis.)

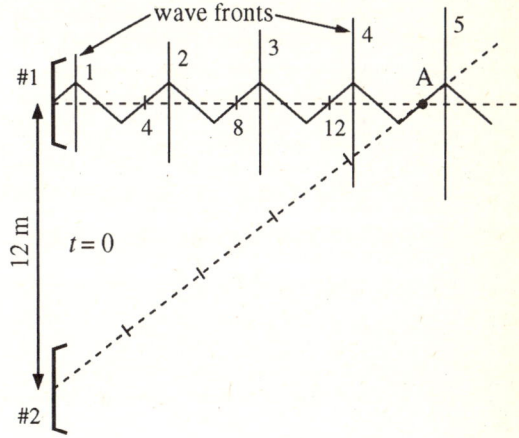

a) What is distance r_2 from A to speaker #2?

b) Draw wave #2 from speaker #2 along the line to point A.

c) Draw and number the wave fronts for wave #2.

d) On the left axes below draw the history graphs showing the displacements D_1 and D_2 of waves #1 and 2 at point A as a function of time for $0 \le t \le 2T$. (The picture shows the situation at $t = 0$.) Then on the right axes draw their superposition, showing the net displacement at point A.

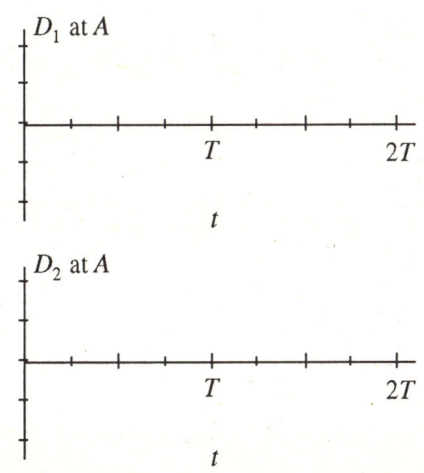
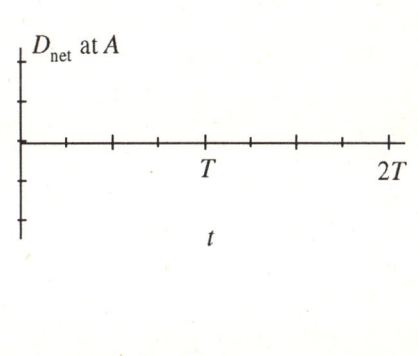

e) Is the interference at point A constructive, destructive, or in between? _____

9. These are the same speakers as in Exercise 8. Now consider point B, distance $r_1 = 9$ m from speaker #1.

a) What is distance r_2 from B to speaker #2?

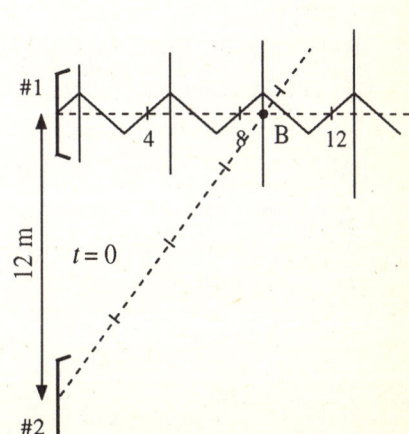

b) Draw wave #2 from speaker #2 along the line to B.

c) Draw the wave fronts for wave #2.

d) On the left axes below draw the history graphs D_1 and D_2 showing the displacement at point B. On the right axes draw their superposition, showing the net displacement at B.

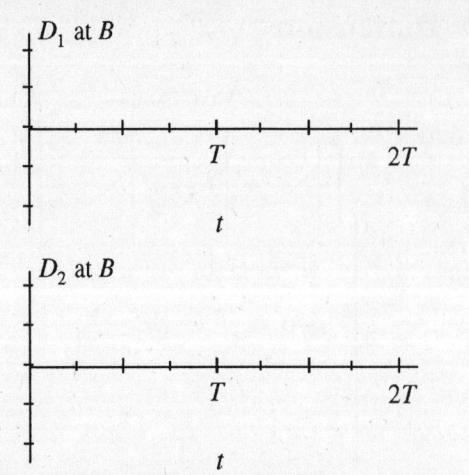
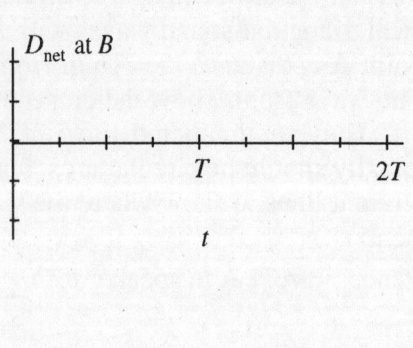

e) Is the interference at point B constructive, destructive, or in between? _____

f) What is the phase ϕ_1 of wave #1 at point B? _____ ϕ_2 of wave #2? _____

What is $\Delta\phi = \phi_2 - \phi_1$? _____

10. Once again consider the two speakers of Exercise 8. The figure shows five points on a line straight out from speaker #1. Their distances r_1 are shown in the table below.

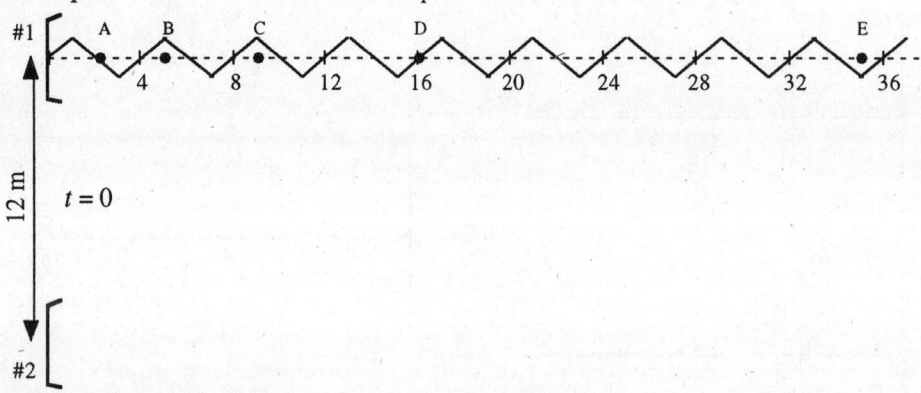

a) For each of these points determine
   i) Distance r_2 (calculated, not just from the picture).
   ii) $\Delta r = r_2 - r_1$.   iii) The ratio $\Delta r/\lambda$.
   iv) The phases ϕ_1 and ϕ_2 of the two waves at $t = 0$, calculated from $\phi = 2\pi(r/\lambda - t/T) + \phi_0$.
   v) The ratio $\Delta\phi/2\pi = |\phi_2 - \phi_1|/2\pi$.
   vi) Whether there is constructive (C) or destructive (D) interference at that point.
Put your answers in the table.

Point	r_1	r_2	Δr	$\Delta r/\lambda$	ϕ_1	ϕ_2	$\Delta\phi/2\pi$	C or D
A	2.2 m							
B	5.0 m							
C	9.0 m							
D	16 m							
E	35 m							

b) Are there any points to the right of E, on the line straight out from speaker #1, for which the interference is either exactly constructive or exactly destructive? If so, where? If not, why not?

c) Suppose you started at speaker #1 and walked straight away from it for 50 m. Describe what you would hear as you walked.

11. The figure shows the wave front pattern emitted by two loudspeakers. Both have $\phi_0 = 0$.

a) Draw a closed circle ● at each point where there is constructive interference. These will be points where two crests overlap *or* two troughs overlap.

b) Draw an open circle o at each point where there is destructive interference. These will be points where a crest overlaps a trough.

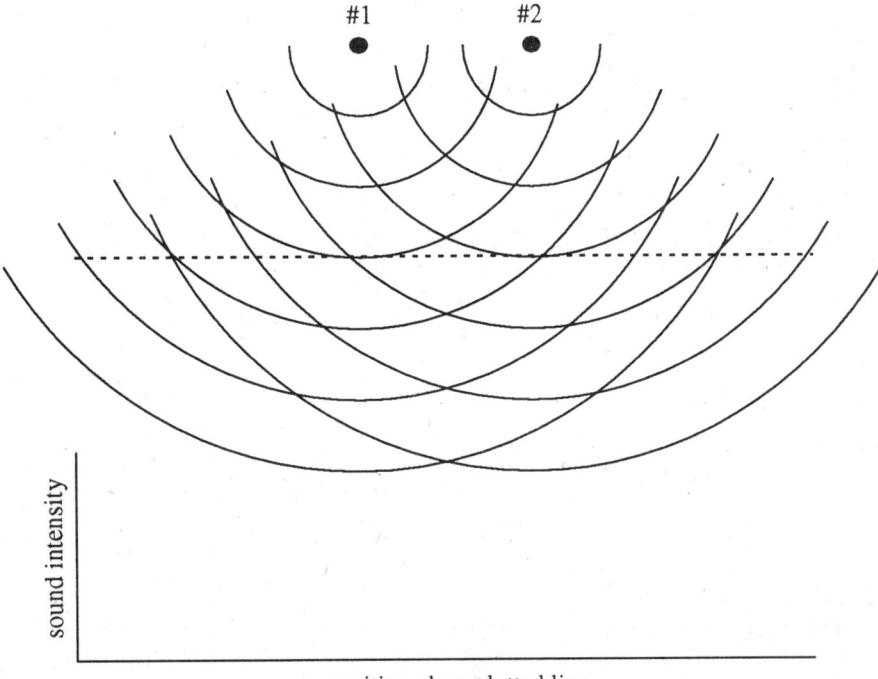

c) Use a black line to connect each "ray" of constructive interference. Use a red line to connect each "ray" of destructive interference.

d) Draw a graph of the sound intensity you would hear if you walked along the horizontal dotted line. Use the same horizontal scale as the figure so that your graph lines up with the figure above it.

e) Suppose the phase constant of speaker #2 is changed to $\phi_0 = \pi$. Describe what would happen to the interference pattern.

16.4 Interference of Light

12. The figure shows a "photograph" of the light intensity falling on a piece of film in an interference experiment. Notice that the light intensity comes "full on" at the edges of each maximum; this is *not* the photograph that would be recorded in Young's double-slit experiment.

a) Draw a graph of light intensity versus position on the film. Make your graph have the same horizontal scale as the "photograph" above it.

b) Is it possible to tell, from the information given, what the wavelength of the light is? If so, what is it? If not, why not?

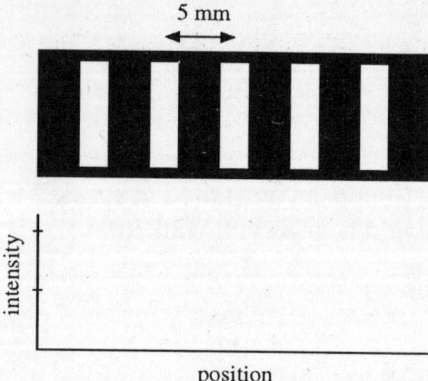

13. The graph represents the light intensity as a function of the position on a screen, as measured in an interference experiment.

a) Draw a "photograph" that would be recorded if a film were put at the position of the screen. Make your "photograph" have the same horizontal scale as the graph above it. Be as accurate as you can. Let the white of the paper be the brightest intensity and a very heavy pencil shading be the darkest.

b) Three positions on the screen are marked as A, B, and C. On the axes below, draw history graphs showing the displacement of the light wave at each of these three positions. Show three cycles, $0 \le t \le 3T$, and use the same vertical scale on all three.

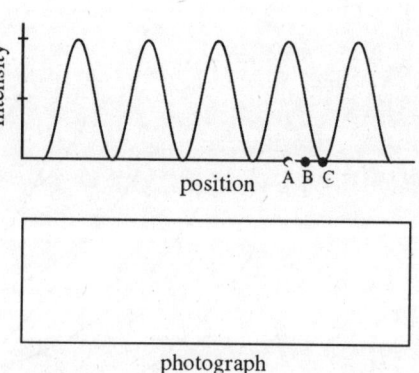

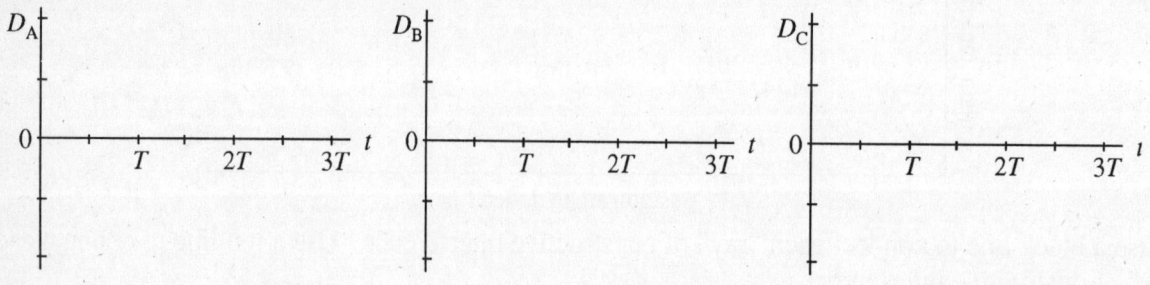

14. In Young's two-slit experiment we usually see the light only on a viewing screen some distance away from the slits. We don't see the light as it propagates through space between the slits and the screen. We can, however, make the light visible with smoke or dust. Consider a two-slit experiment, seen from above, in a smoke-filled room. What would you see as you looked down on the experiment from above? That is, what kind of light and dark pattern of light would you observe in the space between the slits and the screen? Draw the light pattern on the figure below. Label the different parts of your figure as "bright" or "dark."

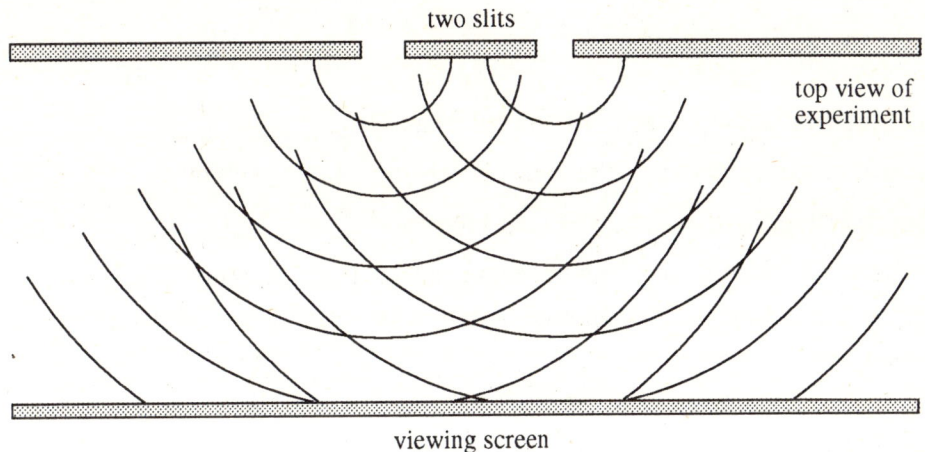

16.5   Interferometers

15. The figure shows a tube through which sound waves with $\lambda = 4$ cm travel from left to right. The wave divides at the first junction and recombines at the second. The dots and triangles show the positions of the wave crests at $t = 0$ — rather like a very simple wave front diagram.

a) Do the recombined waves interfere constructively or destructively? Explain how you can tell.

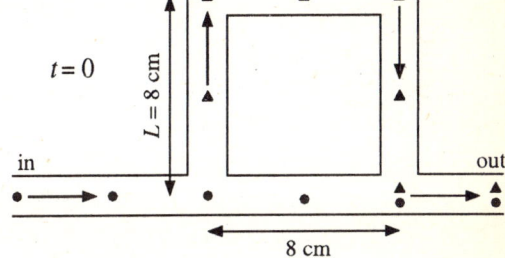

b) How much *extra* distance does the upper wave travel? _____

   How many wavelengths is this extra distance? _____

c) Below are two other tubes with $L = 9$ cm and $L = 10$ cm. Use dots to show the wave crest positions at $t = 0$ for the wave taking the lower path. Use triangles to show the wave crests at $t = 0$ for the wave taking the upper path. The wavelength is $\lambda = 4$ m. Assume that the first crest is at the left edge of the tube, as in the figure above.

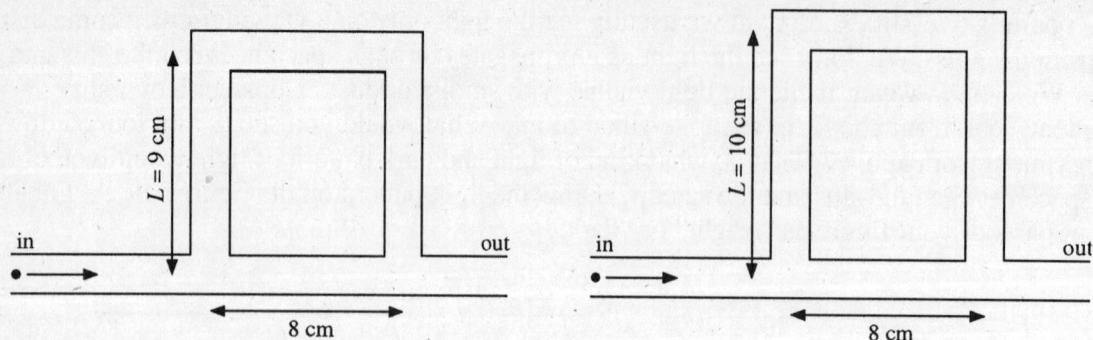

d) What kind of interference does the $L = 9$ cm tube have? _____

   How many *extra* wavelengths does the upper wave travel in this figure? _____

e) What kind of interference does the $L = 10$ cm tube have? _____

   How many *extra* wavelengths does the upper wave travel in this figure? _____

f) Write a general rule stating which values of L give rise to constructive interference.

Chapter 17

Diffraction

17.1 Light in the Shadows

17.2 The Diffraction Grating

1. The figure shows four slits in a diffraction grating. A set of waves is spreading out from each of the slits. Four specific wave paths -- numbered 1 through 4 -- are shown, leaving the slits at angle θ_1. The dotted lines are drawn perpendicular to the paths of the waves.

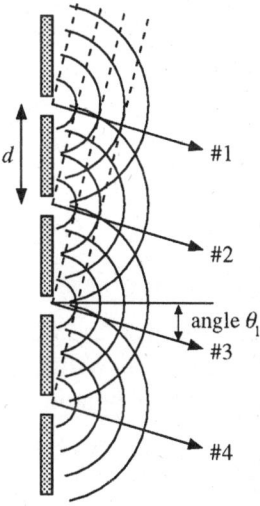

a) Use a colored pencil or heavy shading to show *on the figure* the extra distance traveled by outgoing wave #1 that is *not* traveled by wave #2.

b) How many extra wavelengths does wave #1 travel compared to wave #2? Explain how you can tell from the figure.

c) How many extra wavelengths does wave #2 travel compared to wave #3? _____

d) As these four waves combine some distance away from the grating, will they interfere constructively, destructively, or in between? Explain.

e) Do a geometrical analysis to determine the extra travel distance you identified in a) in terms of the slit spacing d and the angle of travel θ_1. You might want to draw below an enlarged picture of the relevant part of the diagram.

2. a) These are the same slits as in the previous problem, with waves of the same wavelength spreading out on the right side. Draw four paths, originating at the slits, at an angle θ_2 such that the wave along each path travels *two* wavelengths farther than the next wave. Also draw dashed lines at right angles to the travel direction. Your picture should look much like the figure of Exercise 1, but with the waves traveling at a different angle. Use a ruler!

b) Do the same for four paths at angle $\theta_{1/2}$ such that each wave travels *one-half* wavelength farther than the next wave.

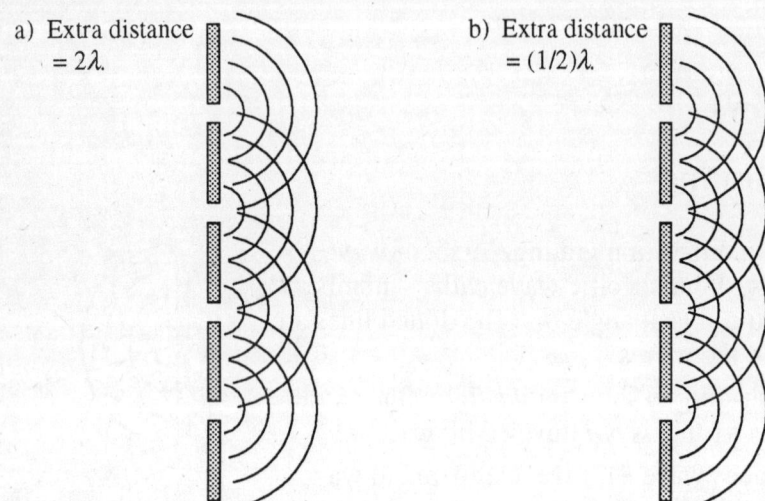

a) Extra distance = 2λ.

b) Extra distance = $(1/2)\lambda$.

3. Suppose the wavelength in Exercise 1 was doubled. (Imagine erasing every-other wave front in the picture.) Would the interference at the same angle θ_1 then be constructive, destructive, or in between? Explain, basing your explanation on an interpretation of the figure, not on some formula.

4. Suppose the slit spacing d in Exercise 1 was doubled while the wavelength and direction θ_1 were unchanged. Would the interference then be constructive, destructive, or in between? Again, base your explanation on the figure.

17.3 Single-Slit Diffraction

5. a) The figure shows graphically the light intensity on a screen behind a 0.2 mm wide slit through which light of 500 nm wavelength is passing. In the space below, draw a *picture* of how a photograph taken at this location would look. Use the same horizontal scale, so that your picture aligns with the graph above. Let the white of the paper represent the brightest intensity and the darkest you can draw with a pencil or pen be the least intensity.

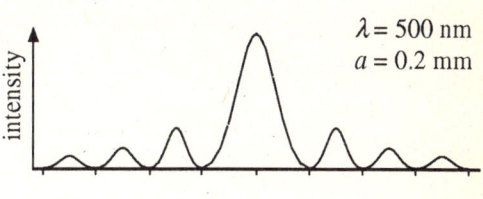

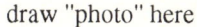

draw "photo" here

b) Using the same horizontal scale as in a), draw graphs showing the diffraction pattern if
  i. $\lambda = 250$ nm, $a = 0.2$ mm
  ii. $\lambda = 1000$ nm, $a = 0.2$ mm
  iii. $\lambda = 500$ nm, $a = 0.1$ mm
  iv. $\lambda = 500$ nm, $a = 0.4$ mm

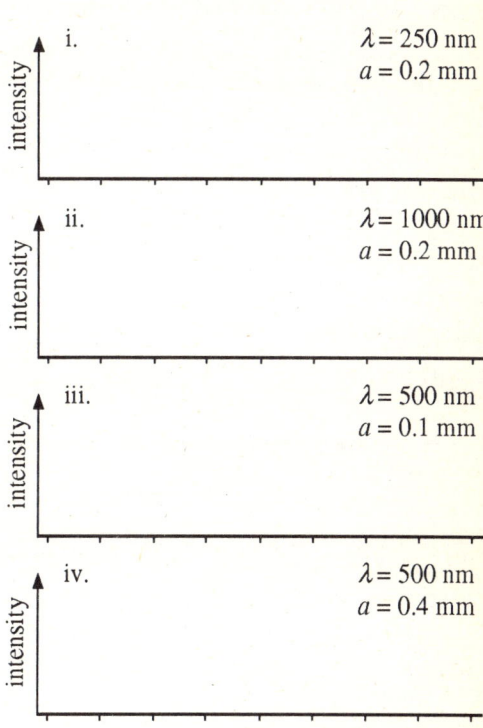

17.4 Circular-Aperture Diffraction

6. The figure shows a "photograph" of a diffraction pattern made by a rectangular opening in a screen. Is the shape of the opening

☐ or ▯ or ▫ ?

Explain your reasoning.

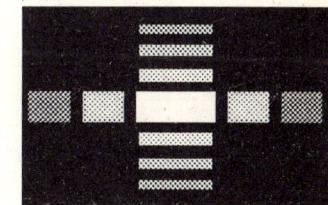

17.5 Diffraction By a Lens

7. For the best possible resolution, to see the finest details, should a microscope illuminate the object with red, white, or blue light? Explain.

17.6 X-Ray Diffraction

8. Below is an atomic lattice with the atoms arranged in what is called a "hexagonal array." One hexagon is outlined to illustrate it. This is a common structure for many crystals.

a) Using a ruler or straight edge, draw lines that start on the atom in the upper left corner and that pass through two or more other <u>equally spaced</u> atoms in the array. These are the atomic planes in the crystal that can diffract X-rays. Number your planes 1, 2, 3, ... with 1 being the horizontal plane extending straight right from the starting atom. You should find a total of 15.

b) Just by eye, or using a protractor, estimate the angle ϕ of each of these atomic planes below the horizontal. Also, count the number of atoms in each plane.

number	angle ϕ	# atoms
___	___	___
___	___	___
___	___	___
___	___	___
___	___	___
___	___	___
___	___	___
___	___	___
___	___	___
___	___	___
___	___	___
___	___	___
___	___	___
___	___	___
___	___	___

Note that angle ϕ is not the angle θ appearing in the Bragg condition. You do, however, need to know ϕ to determine the angles θ for which diffraction occurs.

c) Planes with more atoms will diffract the X-rays more strongly than planes with fewer atoms. Which of your 15 atomic planes have 5 or more atoms? (These are the planes causing most of the x-ray diffraction in this crystal. Diffraction from the other planes would be much weaker.)

Chapter 18

A Closer Look at Light and Matter

No exercises.

DYNAMICS WORKSHEET Name _____ Problem _____

1) Pictorial Model
   a. sketch
   b. coordinate system
   c. symbols for knowns and unknowns

Known information:

Desired unknowns:

2) Physical Model
   a. motion diagram
   b. force identification
   c. free body diagram

3) Mathematical Model

4) Evaluation
   a. sign
   b. units
   c. magnitude

DYNAMICS WORKSHEET   Name _____   Problem _____

1) **Pictorial Model**
   a. sketch
   b. coordinate system
   c. symbols for knowns and unknowns

Known information:
Desired unknowns:

2) **Physical Model**
   a. motion diagram
   b. force identification
   c. free body diagram

3) **Mathematical Model**

4) **Evaluation**
   a. sign
   b. units
   c. magnitude

DYNAMICS WORKSHEET   Name _____   Problem _____

1) Pictorial Model
   a. sketch
   b. coordinate system
   c. symbols for knowns and unknowns

Known information:

Desired unknowns:

2) Physical Model
   a. motion diagram
   b. force identification
   c. free body diagram

3) Mathematical Model

4) Evaluation
   a. sign
   b. units
   c. magnitude

DYNAMICS WORKSHEET   Name _____   Problem _____

1) **Pictorial Model**
   a. sketch
   b. coordinate system
   c. symbols for knowns and unknowns

Known information:

Desired unknowns:

2) **Physical Model**
   a. motion diagram
   b. force identification
   c. free body diagram

3) **Mathematical Model**

4) **Evaluation**
   a. sign
   b. units
   c. magnitude

DYNAMICS WORKSHEET   Name _____ Problem _____

1) Pictorial Model
   a. sketch
   b. coordinate system
   c. symbols for knowns and unknowns

Known information:

Desired unknowns:

2) Physical Model
   a. motion diagram
   b. force identification
   c. free body diagram

3) Mathematical Model

4) Evaluation
   a. sign
   b. units
   c. magnitude

DYNAMICS WORKSHEET   Name _____   Problem _____

1) **Pictorial Model**
   a. sketch
   b. coordinate system
   c. symbols for knowns and unknowns

Known information:

Desired unknowns:

2) **Physical Model**
   a. motion diagram
   b. force identification
   c. free body diagram

3) **Mathematical Model**

4) **Evaluation**
   a. sign
   b. units
   c. magnitude

DYNAMICS WORKSHEET   Name _____   Problem _____

1) **Pictorial Model**
    a. sketch
    b. coordinate system
    c. symbols for knowns and unknowns

Known information:

Desired unknowns:

2) **Physical Model**
    a. motion diagram
    b. force identification
    c. free body diagram

3) **Mathematical Model**

4) **Evaluation**
    a. sign
    b. units
    c. magnitude

DYNAMICS WORKSHEET   Name _____   Problem _____

1) Pictorial Model
   a. sketch
   b. coordinate system
   c. symbols for knowns and unknowns

   Known information:

   Desired unknowns:

2) Physical Model
   a. motion diagram
   b. force identification
   c. free body diagram

3) Mathematical Model

4) Evaluation
   a. sign
   b. units
   c. magnitude

DYNAMICS WORKSHEET   Name _____ Problem _____

1) Pictorial Model
   a. sketch
   b. coordinate system
   c. symbols for knowns and unknowns

Known information:

Desired unknowns:

2) Physical Model
   a. motion diagram
   b. force identification
   c. free body diagram

3) Mathematical Model

4) Evaluation
   a. sign
   b. units
   c. magnitude

DYNAMICS WORKSHEET   Name _____   Problem _____

1) Pictorial Model
   a. sketch
   b. coordinate system
   c. symbols for knowns and unknowns

Known information:

Desired unknowns:

2) Physical Model
   a. motion diagram
   b. force identification
   c. free body diagram

3) Mathematical Model

4) Evaluation
   a. sign
   b. units
   c. magnitude

DYNAMICS WORKSHEET   Name _____   Problem _____

1) Pictorial Model
   a. sketch
   b. coordinate system
   c. symbols for knowns and unknowns

Known information:

Desired unknowns:

2) Physical Model
   a. motion diagram
   b. force identification
   c. free body diagram

3) Mathematical Model

4) Evaluation
   a. sign
   b. units
   c. magnitude

DYNAMICS WORKSHEET   Name _____   Problem _____

1) Pictorial Model
   a. sketch
   b. coordinate system
   c. symbols for knowns and unknowns

Known information:

Desired unknowns:

2) Physical Model
   a. motion diagram
   b. force identification
   c. free body diagram

3) Mathematical Model

4) Evaluation
   a. sign
   b. units
   c. magnitude

DYNAMICS WORKSHEET   Name _____   Problem _____

1) Pictorial Model
   a. sketch
   b. coordinate system
   c. symbols for knowns and unknowns

Known information:

Desired unknowns:

2) Physical Model
   a. motion diagram
   b. force identification
   c. free body diagram

3) Mathematical Model

4) Evaluation
   a. sign
   b. units
   c. magnitude

DYNAMICS WORKSHEET   Name _____   Problem _____

1) **Pictorial Model**
   a. sketch
   b. coordinate system
   c. symbols for knowns and unknowns

   Known information:

   Desired unknowns:

2) **Physical Model**
   a. motion diagram
   b. force identification
   c. free body diagram

3) **Mathematical Model**

4) **Evaluation**
   a. sign
   b. units
   c. magnitude

DYNAMICS WORKSHEET   Name _____   Problem _____

1) **Pictorial Model**
   a. sketch
   b. coordinate system
   c. symbols for knowns and unknowns

Known information:

Desired unknowns:

2) **Physical Model**
   a. motion diagram
   b. force identification
   c. free body diagram

3) **Mathematical Model**

4) **Evaluation**
   a. sign
   b. units
   c. magnitude

DYNAMICS WORKSHEET   Name _____   Problem _____

1) Pictorial Model
   a. sketch
   b. coordinate system
   c. symbols for knowns and unknowns

Known information:

Desired unknowns:

2) Physical Model
   a. motion diagram
   b. force identification
   c. free body diagram

3) Mathematical Model

4) Evaluation
   a. sign
   b. units
   c. magnitude

DYNAMICS WORKSHEET   Name _____   Problem _____

1) Pictorial Model
   a. sketch
   b. coordinate system
   c. symbols for knowns and unknowns

Known information:
Desired unknowns:

2) Physical Model
   a. motion diagram
   b. force identification
   c. free body diagram

3) Mathematical Model

4) Evaluation
   a. sign
   b. units
   c. magnitude

DYNAMICS WORKSHEET   Name _____   Problem _____

1) **Pictorial Model**
   a. sketch
   b. coordinate system
   c. symbols for knowns and unknowns

   Known information:

   Desired unknowns:

2) **Physical Model**
   a. motion diagram
   b. force identification
   c. free body diagram

3) **Mathematical Model**

4) **Evaluation**
   a. sign
   b. units
   c. magnitude

DYNAMICS WORKSHEET   Name _____   Problem _____

1) **Pictorial Model**
   a. sketch
   b. coordinate system
   c. symbols for knowns and unknowns

Known information:
Desired unknowns:

2) **Physical Model**
   a. motion diagram
   b. force identification
   c. free body diagram

3) **Mathematical Model**

4) **Evaluation**
   a. sign
   b. units
   c. magnitude

DYNAMICS WORKSHEET   Name _____ Problem _____

1) Pictorial Model
   a. sketch
   b. coordinate system
   c. symbols for knowns and unknowns

Known information:

Desired unknowns:

2) Physical Model
   a. motion diagram
   b. force identification
   c. free body diagram

3) Mathematical Model

4) Evaluation
   a. sign
   b. units
   c. magnitude

DYNAMICS WORKSHEET   Name _____   Problem _____

1) Pictorial Model
   a. sketch
   b. coordinate system
   c. symbols for knowns and unknowns

Known information:

Desired unknowns:

2) Physical Model
   a. motion diagram
   b. force identification
   c. free body diagram

3) Mathematical Model

4) Evaluation
   a. sign
   b. units
   c. magnitude

DYNAMICS WORKSHEET   Name _____   Problem _____

1) Pictorial Model
   a. sketch
   b. coordinate system
   c. symbols for knowns and unknowns

Known information:
Desired unknowns:

2) Physical Model
   a. motion diagram
   b. force identification
   c. free body diagram

3) Mathematical Model

4) Evaluation
   a. sign
   b. units
   c. magnitude

DYNAMICS WORKSHEET   Name _____   Problem _____

1) Pictorial Model
   a. sketch
   b. coordinate system
   c. symbols for knowns and unknowns

Known information:
Desired unknowns:

2) Physical Model
   a. motion diagram
   b. force identification
   c. free body diagram

3) Mathematical Model

4) Evaluation
   a. sign
   b. units
   c. magnitude

DYNAMICS WORKSHEET Name _____ Problem _____

1) Pictorial Model
   a. sketch
   b. coordinate system
   c. symbols for knowns and unknowns

Known information:
Desired unknowns:

2) Physical Model
   a. motion diagram
   b. force identification
   c. free body diagram

3) Mathematical Model

4) Evaluation
   a. sign
   b. units
   c. magnitude

DYNAMICS WORKSHEET   Name _____   Problem _____

1) **Pictorial Model**
   a. sketch
   b. coordinate system
   c. symbols for knowns and unknowns

Known information:

Desired unknowns:

2) **Physical Model**
   a. motion diagram
   b. force identification
   c. free body diagram

3) **Mathematical Model**

4) **Evaluation**
   a. sign
   b. units
   c. magnitude

DYNAMICS WORKSHEET   Name _____   Problem _____

1) **Pictorial Model**
   a. sketch
   b. coordinate system
   c. symbols for knowns and unknowns

   Known information:

   Desired unknowns:

2) **Physical Model**
   a. motion diagram
   b. force identification
   c. free body diagram

3) **Mathematical Model**

4) **Evaluation**
   a. sign
   b. units
   c. magnitude

DYNAMICS WORKSHEET   Name _____   Problem _____

1) Pictorial Model
   a. sketch
   b. coordinate system
   c. symbols for knowns and unknowns

Known information:
Desired unknowns:

2) Physical Model
   a. motion diagram
   b. force identification
   c. free body diagram

3) Mathematical Model

4) Evaluation
   a. sign
   b. units
   c. magnitude

DYNAMICS WORKSHEET   Name _____   Problem _____

1) **Pictorial Model**
   a. sketch
   b. coordinate system
   c. symbols for knowns and unknowns

Known information:

Desired unknowns:

2) **Physical Model**
   a. motion diagram
   b. force identification
   c. free body diagram

3) **Mathematical Model**

4) **Evaluation**
   a. sign
   b. units
   c. magnitude

DYNAMICS WORKSHEET Name _____ Problem _____

1) Pictorial Model
   a. sketch
   b. coordinate system
   c. symbols for knowns and unknowns

Known information:

Desired unknowns:

2) Physical Model
   a. motion diagram
   b. force identification
   c. free body diagram

3) Mathematical Model

4) Evaluation
   a. sign
   b. units
   c. magnitude

DYNAMICS WORKSHEET   Name _____   Problem _____

1) **Pictorial Model**
   a. sketch
   b. coordinate system
   c. symbols for knowns and unknowns

Known information:
Desired unknowns:

2) **Physical Model**
   a. motion diagram
   b. force identification
   c. free body diagram

3) **Mathematical Model**

4) **Evaluation**
   a. sign
   b. units
   c. magnitude

DYNAMICS WORKSHEET   Name _____   Problem _____

1) **Pictorial Model**
   a. sketch
   b. coordinate system
   c. symbols for knowns and unknowns

   Known information:

   Desired unknowns:

2) **Physical Model**
   a. motion diagram
   b. force identification
   c. free body diagram

3) **Mathematical Model**

4) **Evaluation**
   a. sign
   b. units
   c. magnitude

DYNAMICS WORKSHEET   Name _____   Problem _____

1) Pictorial Model
   a. sketch
   b. coordinate system
   c. symbols for knowns and unknowns

Known information:

Desired unknowns:

2) Physical Model
   a. motion diagram
   b. force identification
   c. free body diagram

3) Mathematical Model

4) Evaluation
   a. sign
   b. units
   c. magnitude

DYNAMICS WORKSHEET   Name _____   Problem _____

1) Pictorial Model
   a. sketch
   b. coordinate system
   c. symbols for knowns and unknowns

Known information:
Desired unknowns:

2) Physical Model
   a. motion diagram
   b. force identification
   c. free body diagram

3) Mathematical Model

4) Evaluation
   a. sign
   b. units
   c. magnitude

DYNAMICS WORKSHEET Name _____ Problem _____

1) Pictorial Model
   a. sketch
   b. coordinate system
   c. symbols for knowns and unknowns

Known information:

Desired unknowns:

2) Physical Model
   a. motion diagram
   b. force identification
   c. free body diagram

3) Mathematical Model

4) Evaluation
   a. sign
   b. units
   c. magnitude

DYNAMICS WORKSHEET   Name _____   Problem _____

1) Pictorial Model
   a. sketch
   b. coordinate system
   c. symbols for knowns and unknowns

Known information:

Desired unknowns:

2) Physical Model
   a. motion diagram
   b. force identification
   c. free body diagram

3) Mathematical Model

4) Evaluation
   a. sign
   b. units
   c. magnitude

DYNAMICS WORKSHEET   Name _____   Problem _____

1) **Pictorial Model**
   a. sketch
   b. coordinate system
   c. symbols for knowns and unknowns

   Known information:

   Desired unknowns:

2) **Physical Model**
   a. motion diagram
   b. force identification
   c. free body diagram

3) **Mathematical Model**

4) **Evaluation**
   a. sign
   b. units
   c. magnitude

DYNAMICS WORKSHEET Name _____ Problem _____

1) Pictorial Model
   a. sketch
   b. coordinate system
   c. symbols for knowns and unknowns

Known information:

Desired unknowns:

2) Physical Model
   a. motion diagram
   b. force identification
   c. free body diagram

3) Mathematical Model

4) Evaluation
   a. sign
   b. units
   c. magnitude

DYNAMICS WORKSHEET   Name _____   Problem _____

1) Pictorial Model
   a. sketch
   b. coordinate system
   c. symbols for knowns and unknowns

Known information:

Desired unknowns:

2) Physical Model
   a. motion diagram
   b. force identification
   c. free body diagram

3) Mathematical Model

4) Evaluation
   a. sign
   b. units
   c. magnitude

DYNAMICS WORKSHEET Name _____ Problem _____

1) Pictorial Model
   a. sketch
   b. coordinate system
   c. symbols for knowns and unknowns

Known information:
Desired unknowns:

2) Physical Model
   a. motion diagram
   b. force identification
   c. free body diagram

3) Mathematical Model

4) Evaluation
   a. sign
   b. units
   c. magnitude

DYNAMICS WORKSHEET   Name _____   Problem _____

1) Pictorial Model
   a. sketch
   b. coordinate system
   c. symbols for knowns and unknowns

Known information:

Desired unknowns:

2) Physical Model
   a. motion diagram
   b. force identification
   c. free body diagram

3) Mathematical Model

4) Evaluation
   a. sign
   b. units
   c. magnitude

DYNAMICS WORKSHEET Name _____ Problem _____

1) Pictorial Model
   a. sketch
   b. coordinate system
   c. symbols for knowns and unknowns

Known information:
Desired unknowns:

2) Physical Model
   a. motion diagram
   b. force identification
   c. free body diagram

3) Mathematical Model

4) Evaluation
   a. sign
   b. units
   c. magnitude

DYNAMICS WORKSHEET   Name _____ Problem _____

1) **Pictorial Model**
   a. sketch
   b. coordinate system
   c. symbols for knowns and unknowns

Known information:

Desired unknowns:

2) **Physical Model**
   a. motion diagram
   b. force identification
   c. free body diagram

3) **Mathematical Model**

4) **Evaluation**
   a. sign
   b. units
   c. magnitude

DYNAMICS WORKSHEET   Name _____   Problem _____

1) **Pictorial Model**
   a. sketch
   b. coordinate system
   c. symbols for knowns and unknowns

Known information:
Desired unknowns:

2) **Physical Model**
   a. motion diagram
   b. force identification
   c. free body diagram

3) **Mathematical Model**

4) **Evaluation**
   a. sign
   b. units
   c. magnitude

DYNAMICS WORKSHEET   Name _____   Problem _____

1) Pictorial Model
   a. sketch
   b. coordinate system
   c. symbols for knowns and unknowns

Known information:

Desired unknowns:

2) Physical Model
   a. motion diagram
   b. force identification
   c. free body diagram

3) Mathematical Model

4) Evaluation
   a. sign
   b. units
   c. magnitude

DYNAMICS WORKSHEET   Name _____   Problem _____

1) Pictorial Model
   a. sketch
   b. coordinate system
   c. symbols for knowns and unknowns

   Known information:

   Desired unknowns:

2) Physical Model
   a. motion diagram
   b. force identification
   c. free body diagram

3) Mathematical Model

4) Evaluation
   a. sign
   b. units
   c. magnitude

DYNAMICS WORKSHEET   Name _____   Problem _____

1) Pictorial Model
   a. sketch
   b. coordinate system
   c. symbols for knowns and unknowns

Known information:
Desired unknowns:

2) Physical Model
   a. motion diagram
   b. force identification
   c. free body diagram

3) Mathematical Model

4) Evaluation
   a. sign
   b. units
   c. magnitude

DYNAMICS WORKSHEET   Name _____   Problem _____

1) **Pictorial Model**
   a. sketch
   b. coordinate system
   c. symbols for knowns and unknowns

   Known information:

   Desired unknowns:

2) **Physical Model**
   a. motion diagram
   b. force identification
   c. free body diagram

3) **Mathematical Model**

4) **Evaluation**
   a. sign
   b. units
   c. magnitude

DYNAMICS WORKSHEET   Name _____   Problem _____

1) Pictorial Model
   a. sketch
   b. coordinate system
   c. symbols for knowns and unknowns

Known information:

Desired unknowns:

2) Physical Model
   a. motion diagram
   b. force identification
   c. free body diagram

3) Mathematical Model

4) Evaluation
   a. sign
   b. units
   c. magnitude

DYNAMICS WORKSHEET   Name _____   Problem _____

1) Pictorial Model
   a. sketch
   b. coordinate system
   c. symbols for knowns and unknowns

Known information:

Desired unknowns:

2) Physical Model
   a. motion diagram
   b. force identification
   c. free body diagram

3) Mathematical Model

4) Evaluation
   a. sign
   b. units
   c. magnitude

DYNAMICS WORKSHEET   Name _____   Problem _____

1) Pictorial Model
   a. sketch
   b. coordinate system
   c. symbols for knowns and unknowns

Known information:

Desired unknowns:

2) Physical Model
   a. motion diagram
   b. force identification
   c. free body diagram

3) Mathematical Model

4) Evaluation
   a. sign
   b. units
   c. magnitude

CONSERVATION WORKSHEET Name _____ Problem _____

1) **Pictorial Model**
   a. sketch of before and after
   b. coordinate system
   c. symbols for knowns and unknowns

Known information:

Desired unknowns:

2) **Energy Balance**

enter +, –, or 0	Part 1	(if needed) Part 2
kinetic energy	_____	_____
grav. pot. eng.	_____	_____
spring pot. eng.	_____	_____
internal energy	_____	_____

3) **Work and Impluse**
   Identify forces for which you need to calculate work or impluse. Do it!

4) **Mathematical Model**

5) **Evaluation**
   a. sign
   b. units
   c. magnitude

CONSERVATION WORKSHEET   Name _____   Problem _____

1) **Pictorial Model**
   a. sketch of before and after
   b. coordinate system
   c. symbols for knowns and unknowns

Known information:

Desired unknowns:

2) **Energy Balance**

enter +, –, or 0	Part 1	(if needed) Part 2
kinetic energy	_____	_____
grav. pot. eng.	_____	_____
spring pot. eng.	_____	_____
internal energy	_____	_____

3) **Work and Impluse**
   Identify forces for which you need to calculate work or impluse. Do it!

4) **Mathematical Model**

5) **Evaluation**
   a. sign
   b. units
   c. magnitude

CONSERVATION WORKSHEET   Name _____   Problem _____

1) **Pictorial Model**
   a. sketch of before and after
   b. coordinate system
   c. symbols for knowns and unknowns

   Known information:

   Desired unknowns:

2) **Energy Balance**

enter +, –, or 0	Part 1	(if needed) Part 2
kinetic energy	_____	_____
grav. pot. eng.	_____	_____
spring pot. eng.	_____	_____
internal energy	_____	_____

3) **Work and Impulse**
   Identify forces for which you need to calculate work or impulse. Do it!

4) **Mathematical Model**

5) **Evaluation**
   a. sign
   b. units
   c. magnitude

CONSERVATION WORKSHEET   Name _____   Problem _____

1) **Pictorial Model**
    a. sketch of before and after
    b. coordinate system
    c. symbols for knowns and unknowns

   Known information:

   Desired unknowns:

2) **Energy Balance**

enter +, −, or 0	Part 1	(if needed) Part 2
kinetic energy	_____	_____
grav. pot. eng.	_____	_____
spring pot. eng.	_____	_____
internal energy	_____	_____

3) **Work and Impluse**
   Identify forces for which you need to calculate work or impluse. Do it!

4) **Mathematical Model**

5) **Evaluation**
    a. sign
    b. units
    c. magnitude

CONSERVATION WORKSHEET   Name _____ Problem _____

1) Pictorial Model
  a. sketch of before and after
  b. coordinate system
  c. symbols for knowns and unknowns

Known information:
Desired unknowns:

2) Energy Balance

enter +, –, or 0   Part 1   (if needed) Part 2

kinetic energy   _____   _____

grav. pot. eng.   _____   _____

spring pot. eng.   _____   _____

internal energy   _____   _____

3) Work and Impulse
  Identify forces for which you need to calculate work or impulse. Do it!

4) Mathematical Model

5) Evaluation
  a. sign
  b. units
  c. magnitude

CONSERVATION WORKSHEET   Name _____   Problem _____

1) **Pictorial Model**
   a. sketch of before and after
   b. coordinate system
   c. symbols for knowns and unknowns

Known information:

Desired unknowns:

2) **Energy Balance**

enter +, –, or 0	Part 1	(if needed) Part 2
kinetic energy	_____	_____
grav. pot. eng.	_____	_____
spring pot. eng.	_____	_____
internal energy	_____	_____

3) **Work and Impulse**
   Identify forces for which you need to calculate work or impulse. Do it!

4) **Mathematical Model**

5) **Evaluation**
   a. sign
   b. units
   c. magnitude

CONSERVATION WORKSHEET   Name _____   Problem _____

● 1) Pictorial Model
    a. sketch of before and after
    b. coordinate system
    c. symbols for knowns and unknowns

Known information:
Desired unknowns:

2) Energy Balance

    enter +, –, or 0   Part 1   (if needed) Part 2

    kinetic energy   _____   _____
    grav. pot. eng.   _____   _____
    spring pot. eng.   _____   _____
    internal energy   _____   _____

3) Work and Impulse
    Identify forces for which you need to calculate work or impulse. Do it!

● 4) Mathematical Model

● 5) Evaluation
    a. sign
    b. units
    c. magnitude

CONSERVATION WORKSHEET   Name _____   Problem _____

1) Pictorial Model
   a. sketch of before and after
   b. coordinate system
   c. symbols for knowns and unknowns

Known information:

Desired unknowns:

2) Energy Balance

enter +, –, or 0	Part 1	(if needed) Part 2
kinetic energy	_____	_____
grav. pot. eng.	_____	_____
spring pot. eng.	_____	_____
internal energy	_____	_____

3) Work and Impulse
   Identify forces for which you need to calculate work or impluse. Do it!

4) Mathematical Model

5) Evaluation
   a. sign
   b. units
   c. magnitude

CONSERVATION WORKSHEET   Name _____   Problem _____

1) Pictorial Model
   a. sketch of before and after
   b. coordinate system
   c. symbols for knowns and unknowns

Known information:

Desired unknowns:

2) Energy Balance

enter +, –, or 0	Part 1	(if needed) Part 2
kinetic energy	_____	_____
grav. pot. eng.	_____	_____
spring pot. eng.	_____	_____
internal energy	_____	_____

3) Work and Impulse
   Identify forces for which you need to calculate work or impulse. Do it!

4) Mathematical Model

5) Evaluation
   a. sign
   b. units
   c. magnitude

CONSERVATION WORKSHEET   Name _____   Problem _____

1) Pictorial Model
   a. sketch of before and after
   b. coordinate system
   c. symbols for knowns and unknowns

Known information:

Desired unknowns:

2) Energy Balance

enter +, –, or 0	Part 1	(if needed) Part 2
kinetic energy	_____	_____
grav. pot. eng.	_____	_____
spring pot. eng.	_____	_____
internal energy	_____	_____

3) Work and Impluse
Identify forces for which you need to calculate work or impulse. Do it!

4) Mathematical Model

5) Evaluation
   a. sign
   b. units
   c. magnitude

CONSERVATION WORKSHEET Name _____ Problem _____

1) Pictorial Model
   a. sketch of before and after
   b. coordinate system
   c. symbols for knowns and unknowns

Known information:

Desired unknowns:

2) Energy Balance

enter +, –, or 0	Part 1	(if needed) Part 2
kinetic energy	_____	_____
grav. pot. eng.	_____	_____
spring pot. eng.	_____	_____
internal energy	_____	_____

3) Work and Impulse
   Identify forces for which you need to calculate work or impulse. Do it!

4) Mathematical Model

5) Evaluation
   a. sign
   b. units
   c. magnitude

CONSERVATION WORKSHEET Name _____ Problem _____

1) Pictorial Model
   a. sketch of before and after
   b. coordinate system
   c. symbols for knowns and unknowns

Known information:

Desired unknowns:

2) Energy Balance

enter +, –, or 0	Part 1	(if needed) Part 2
kinetic energy	_____	_____
grav. pot. eng.	_____	_____
spring pot. eng.	_____	_____
internal energy	_____	_____

3) Work and Impulse
   Identify forces for which you need to calculate work or impulse. Do it!

4) Mathematical Model

5) Evaluation
   a. sign
   b. units
   c. magnitude

CONSERVATION WORKSHEET   Name _____   Problem _____

1) Pictorial Model
    a. sketch of before and after
    b. coordinate system
    c. symbols for knowns and unknowns

Known information:

Desired unknowns:

2) Energy Balance

enter +, –, or 0	Part 1	(if needed) Part 2
kinetic energy	_____	_____
grav. pot. eng.	_____	_____
spring pot. eng.	_____	_____
internal energy	_____	_____

3) Work and Impluse
Identify forces for which you need to calculate work or impluse. Do it!

4) Mathematical Model

5) Evaluation
    a. sign
    b. units
    c. magnitude

CONSERVATION WORKSHEET   Name _____   Problem _____

1) **Pictorial Model**
   a. sketch of before and after
   b. coordinate system
   c. symbols for knowns and unknowns

   Known information:

   Desired unknowns:

2) **Energy Balance**

   enter +, –, or 0   Part 1   (if needed) Part 2

   kinetic energy   _____   _____

   grav. pot. eng.   _____   _____

   spring pot. eng.   _____   _____

   internal energy   _____   _____

3) **Work and Impulse**
   Identify forces for which you need to calculate work or impluse. Do it!

4) **Mathematical Model**

5) **Evaluation**
   a. sign
   b. units
   c. magnitude

CONSERVATION WORKSHEET   Name _____   Problem _____

1) **Pictorial Model**
   a. sketch of before and after
   b. coordinate system
   c. symbols for knowns and unknowns

   Known information:

   Desired unknowns:

2) **Energy Balance**

	Part 1	(if needed) Part 2
enter +, –, or 0		
kinetic energy	_____	_____
grav. pot. eng.	_____	_____
spring pot. eng.	_____	_____
internal energy	_____	_____

3) **Work and Impluse**
   Identify forces for which you need to calculate work or impluse. Do it!

4) **Mathematical Model**

5) **Evaluation**
   a. sign
   b. units
   c. magnitude

CONSERVATION WORKSHEET Name _____ Problem _____

1) Pictorial Model
   a. sketch of before and after
   b. coordinate system
   c. symbols for knowns and unknowns

Known information:

Desired unknowns:

2) Energy Balance

enter +, –, or 0	Part 1	(if needed) Part 2
kinetic energy	_____	_____
grav. pot. eng.	_____	_____
spring pot. eng.	_____	_____
internal energy	_____	_____

3) Work and Impulse
   Identify forces for which you need to calculate work or impulse. Do it!

4) Mathematical Model

5) Evaluation
   a. sign
   b. units
   c. magnitude

CONSERVATION WORKSHEET   Name _____   Problem _____

1) **Pictorial Model**
   a. sketch of before and after
   b. coordinate system
   c. symbols for knowns and unknowns

Known information:

Desired unknowns:

2) **Energy Balance**

enter +, −, or 0	Part 1	(if needed) Part 2
kinetic energy	_____	_____
grav. pot. eng.	_____	_____
spring pot. eng.	_____	_____
internal energy	_____	_____

3) **Work and Impluse**
   Identify forces for which you need to calculate work or impulse. Do it!

4) **Mathematical Model**

5) **Evaluation**
   a. sign
   b. units
   c. magnitude

CONSERVATION WORKSHEET Name _____ Problem _____

1) **Pictorial Model**
   a. sketch of before and after
   b. coordinate system
   c. symbols for knowns and unknowns

   Known information:

   Desired unknowns:

2) **Energy Balance**

enter +, −, or 0	Part 1	(if needed) Part 2
kinetic energy	_____	_____
grav. pot. eng.	_____	_____
spring pot. eng.	_____	_____
internal energy	_____	_____

3) **Work and Impulse**
   Identify forces for which you need to calculate work or impulse. Do it!

4) **Mathematical Model**

5) **Evaluation**
   a. sign
   b. units
   c. magnitude

CONSERVATION WORKSHEET   Name _____   Problem _____

1) **Pictorial Model**
   a. sketch of before and after
   b. coordinate system
   c. symbols for knowns and unknowns

Known information:
Desired unknowns:

2) **Energy Balance**

enter +, –, or 0	Part 1	(if needed) Part 2
kinetic energy	_____	_____
grav. pot. eng.	_____	_____
spring pot. eng.	_____	_____
internal energy	_____	_____

3) **Work and Impulse**
   Identify forces for which you need to calculate work or impulse. Do it!

4) **Mathematical Model**

5) **Evaluation**
   a. sign
   b. units
   c. magnitude

CONSERVATION WORKSHEET   Name _____   Problem _____

1) Pictorial Model
   a. sketch of before and after
   b. coordinate system
   c. symbols for knowns and unknowns

Known information:
Desired unknowns:

2) Energy Balance

enter +, −, or 0	Part 1	(if needed) Part 2
kinetic energy	_____	_____
grav. pot. eng.	_____	_____
spring pot. eng.	_____	_____
internal energy	_____	_____

3) Work and Impulse
   Identify forces for which you need to calculate work or impulse. Do it!

4) Mathematical Model

5) Evaluation
   a. sign
   b. units
   c. magnitude

CONSERVATION WORKSHEET Name _____ Problem _____

1) **Pictorial Model**
   a. sketch of before and after
   b. coordinate system
   c. symbols for knowns and unknowns

   ┌─────────────────────────────┐
   │ Known information: │
   │   │
   │   │
   │   │
   │   │
   │   │
   │   │
   │ Desired unknowns: │
   └─────────────────────────────┘

2) **Energy Balance**

   enter +, −, or 0   Part 1   (if needed) Part 2

   kinetic energy   _____   _____
   grav. pot. eng.   _____   _____
   spring pot. eng.   _____   _____
   internal energy   _____   _____

3) **Work and Impluse**
   Identify forces for which you need to calculate work or impluse. Do it!

4) **Mathematical Model**

5) **Evaluation**
   a. sign
   b. units
   c. magnitude

CONSERVATION WORKSHEET   Name _____   Problem _____

1) Pictorial Model
   a. sketch of before and after
   b. coordinate system
   c. symbols for knowns and unknowns

Known information:

Desired unknowns:

2) Energy Balance

enter +, –, or 0	Part 1	(if needed) Part 2
kinetic energy	_____	_____
grav. pot. eng.	_____	_____
spring pot. eng.	_____	_____
internal energy	_____	_____

3) Work and Impluse
Identify forces for which you need to calculate work or impluse. Do it!

4) Mathematical Model

5) Evaluation
   a. sign
   b. units
   c. magnitude

CONSERVATION WORKSHEET   Name _____   Problem _____

1) **Pictorial Model**
   a. sketch of before and after
   b. coordinate system
   c. symbols for knowns and unknowns

Known information:

Desired unknowns:

2) **Energy Balance**

enter +, –, or 0	Part 1	(if needed) Part 2
kinetic energy	_____	_____
grav. pot. eng.	_____	_____
spring pot. eng.	_____	_____
internal energy	_____	_____

3) **Work and Impluse**
   Identify forces for which you need to calculate work or impluse. Do it!

4) **Mathematical Model**

5) **Evaluation**
   a. sign
   b. units
   c. magnitude

CONSERVATION WORKSHEET   Name _____   Problem _____

1) **Pictorial Model**
   a. sketch of before and after
   b. coordinate system
   c. symbols for knowns and unknowns

Known information:

Desired unknowns:

2) **Energy Balance**

enter +, −, or 0	Part 1	(if needed) Part 2
kinetic energy	_____	_____
grav. pot. eng.	_____	_____
spring pot. eng.	_____	_____
internal energy	_____	_____

3) **Work and Impulse**
   Identify forces for which you need to calculate work or impulse. Do it!

4) **Mathematical Model**

5) **Evaluation**
   a. sign
   b. units
   c. magnitude

CONSERVATION WORKSHEET Name _____ Problem _____

● 1) **Pictorial Model**
    a. sketch of before and after
    b. coordinate system
    c. symbols for knowns and unknowns

Known information:
Desired unknowns:

2) **Energy Balance**

enter +, −, or 0	Part 1	(if needed) Part 2
kinetic energy	_____	_____
grav. pot. eng.	_____	_____
spring pot. eng.	_____	_____
internal energy	_____	_____

3) **Work and Impluse**
    Identify forces for which you need to calculate work or impluse. Do it!

● 4) **Mathematical Model**

● 5) **Evaluation**
    a. sign
    b. units
    c. magnitude

CONSERVATION WORKSHEET Name _____ Problem _____

1) **Pictorial Model**
   a. sketch of before and after
   b. coordinate system
   c. symbols for knowns and unknowns

Known information:

Desired unknowns:

2) **Energy Balance**

enter +, −, or 0	Part 1	(if needed) Part 2
kinetic energy	_____	_____
grav. pot. eng.	_____	_____
spring pot. eng.	_____	_____
internal energy	_____	_____

3) **Work and Impulse**
   Identify forces for which you need to calculate work or impluse. Do it!

4) **Mathematical Model**

5) **Evaluation**
   a. sign
   b. units
   c. magnitude

CONSERVATION WORKSHEET   Name _____   Problem _____

1) **Pictorial Model**
   a. sketch of before and after
   b. coordinate system
   c. symbols for knowns and unknowns

   Known information:

   Desired unknowns:

2) **Energy Balance**

enter +, −, or 0	Part 1	(if needed) Part 2
kinetic energy	_____	_____
grav. pot. eng.	_____	_____
spring pot. eng.	_____	_____
internal energy	_____	_____

3) **Work and Impulse**
   Identify forces for which you need to calculate work or impulse. Do it!

4) **Mathematical Model**

5) **Evaluation**
   a. sign
   b. units
   c. magnitude

CONSERVATION WORKSHEET Name _____ Problem _____

1) **Pictorial Model**
   a. sketch of before and after
   b. coordinate system
   c. symbols for knowns and unknowns

   Known information:

   Desired unknowns:

2) **Energy Balance**

enter +, −, or 0	Part 1	(if needed) Part 2
kinetic energy	_____	_____
grav. pot. eng.	_____	_____
spring pot. eng.	_____	_____
internal energy	_____	_____

3) **Work and Impluse**
   Identify forces for which you need to calculate work or impluse. Do it!

4) **Mathematical Model**

5) **Evaluation**
   a. sign
   b. units
   c. magnitude

CONSERVATION WORKSHEET Name _____ Problem _____

1) Pictorial Model
   a. sketch of before and after
   b. coordinate system
   c. symbols for knowns and unknowns

Known information:

Desired unknowns:

2) Energy Balance

enter +, –, or 0	Part 1	(if needed) Part 2
kinetic energy	_____	_____
grav. pot. eng.	_____	_____
spring pot. eng.	_____	_____
internal energy	_____	_____

3) Work and Impulse
   Identify forces for which you need to calculate work or impluse. Do it!

4) Mathematical Model

5) Evaluation
   a. sign
   b. units
   c. magnitude

CONSERVATION WORKSHEET Name _____ Problem _____

1) **Pictorial Model**
   a. sketch of before and after
   b. coordinate system
   c. symbols for knowns and unknowns

Known information:

Desired unknowns:

2) **Energy Balance**

enter +, –, or 0	Part 1	(if needed) Part 2
kinetic energy	_____	_____
grav. pot. eng.	_____	_____
spring pot. eng.	_____	_____
internal energy	_____	_____

3) **Work and Impluse**
   Identify forces for which you need to calculate work or impluse. Do it!

4) **Mathematical Model**

5) **Evaluation**
   a. sign
   b. units
   c. magnitude